AF770669

Methods in Cell Biology

VOLUME 55

Laser Tweezers in Cell Biology

ASCB

Series Editors

Leslie Wilson
Department of Biological Sciences
University of California, Santa Barbara
Santa Barbara, California

Paul Matsudaira
Whitehead Institute for Biomedical Research and
Department of Biology
Massachusetts Institute of Technology
Cambridge, Massachusetts

QH
585
.M47
Vol. 55

Methods in Cell Biology

Prepared under the Auspices of the American Society for Cell Biology

VOLUME 55

Laser Tweezers in Cell Biology

Edited by

Michael P. Sheetz
Department of Cell Biology
Duke University Medical Center
Durham, North Carolina

UTD LIBRARY
2601 N. FLOYD RD.
RICHARDSON, TX 75083
WITHDRAWN

ACV 6292

ACADEMIC PRESS
San Diego London Boston New York Sydney Tokyo Toronto

UT-DALLAS LIBRARIES
3 1863 004 961 904

Cover photograph (paperback edition only) provided by Dan P. Felsenfeld.

This book is printed on acid-free paper. ♾

Copyright © 1998 by ACADEMIC PRESS

All Rights Reserved.
No part of this publication may be reproduced or transmitted in any form or by any means, electronic or mechanical, including photocopy, recording, or any information storage and retrieval system, without permission in writing from the Publisher.
The appearance of the code at the bottom of the first page of a chapter in this book indicates the Publisher's consent that copies of the chapter may be made for personal or internal use of specific clients. This consent is given on the condition, however, that the copier pay the stated per copy fee through the Copyright Clearance Center, Inc. (222 Rosewood Drive, Danvers, Massachusetts 01923), for copying beyond that permitted by Sections 107 or 108 of the U.S. Copyright Law. This consent does not extend to other kinds of copying, such as copying for general distribution, for advertising or promotional purposes, for creating new collective works, or for resale. Copy fees for pre-1997 chapters are as shown on the title pages. If no fee code appears on the title page, the copy fee is the same as for current chapters.
0091-679X/97 $25.00

Academic Press
a division of Harcourt Brace & Company
525 B Street, Suite 1900, San Diego, California 92101-4495, USA
http://www.apnet.com

Academic Press Limited
24-28 Oval Road, London NW1 7DX, UK
http://www.hbuk.co.uk/ap/

International Standard Book Number: 0-12-639370-2 (comb-bound)

International Standard Book Number: 0-12-564157-5 (case)

PRINTED IN THE UNITED STATES OF AMERICA
97 98 99 00 01 02 EB 9 8 7 6 5 4 3 2 1

CONTENTS

CONTRIBUTORS

Numbers in parentheses indicate the pages on which the authors' contributions begin.

A. Ashkin (1), AT&T Bell Laboratories, Holmdel, New Jersey 07733

Michael W. Berns (71), Beckman Laser Institute and Medical Clinic, University of California at Irvine, Irvine, California 92612

Jianwu Dai (157), Department of Cell Biology, Duke University Medical Center, Durham, North Carolina 27710

Harald Felgner (195), Adolf-Butenandt-Institut, Zellbiologie, University of Munich, 80336 Munich, Germany

Jeffrey T. Finer (47), Department of Biochemistry, Stanford University School of Medicine, Stanford, California 94305

Takahiro Fujiwara (173), Department of Life Sciences, Graduate School of Arts and Sciences, The University of Tokyo, Komaba 3-8-1, Meguro-ku, Tokyo 153, Japan

Takashi Funatsu (117), Yanagida BioMotron Project, ERATO, ST, Senba-higasi 2-4-14, Mino, Osaka, 562 Japan

Frederick Gittes (129), Department of Physics, and Biophysics Research Division, University of Michigan, Ann Arbor, Michigan 48109

Franz Grolig (195), Institut für Allgemeine Botanik und Pflauzeuphysiologie, University of Giessen, 35390 Giessen, Germany

Yoshie Harada (117), Yanagida BioMotron Project, ERATO, JST, Senba-higasi 2-4-14, Mino, Osaka, 562 Japan

Hideo Higuchi (117), Yanagida BioMotron Project, ERATO, JST, Senba-higasi 2-4-14, Mino, Osaka, 562 Japan

Yoshiharu Ishii (117), Yanagida BioMotron Project, ERATO, JST, Senba-higasi 2-4-14, Mino, Osaka, 562 Japan

Scot C. Kuo (43), Department of Biomedical Engineering, The Johns Hopkins University, Baltimore, Maryland 21205

Akihiro Kusumi (173), Department of Biological Science, Graduate School of Science, Nagoya University, Nagoya 464-01, Japan

Hong Liang (71), Beckman Laser Institute and Medical Clinic, University of California at Irvine, Irvine, California 92612

Amit D. Mehta (47), Department of Biochemistry, Stanford University School of Medicine, Stanford, California 94305

Justin E. Molloy (205), Department of Biology, University of York, York, YO1 5DD, United Kingdom

Otto Müller (195), Adolf-Butenandt-Institut, Zellbiologie, University of Munich, 80336 Munich, Germany

Kiwamu Saito (117), Yanagida BioMotron Project, ERATO, JST, Senba-higasi 2-4-14, Mino, Osaka, 562 Japan

Yasushi Sako (173), Department of Biological Sciences, Graduate School of Science, Nagoya University, Nagoya 464-01, Japan

Manfred Schliwa (195), Adolf-Butenandt-Institut, Zellbiologie, University of Munich, 80336 Munich, Germany

Christoph F. Schmidt (129), Department of Physics, and Biophysics Research Division, University of Michigan, Ann Arbor, Michigan 48109

Michael P. Sheetz (29, 157), Department of Cell Biology, Duke University Medical Center, Durham, North Carolina 27710

James A. Spudich (47), Department of Biochemistry, Stanford University School of Medicine, Stanford, California 94305

Ronald E. Sterba (29), Department of Cell Biology, Duke University Medical Center, Durham, North Carolina 27710

Andrea L. Stout[1] (99), School of Applied and Engineering Physics, Cornell University, Ithaca, New York 14853

Yona Tadir (71), Beckman Laser Institute and Medical Clinic, University of California at Irvine, Irvine, California 92612

Makio Tokunaga (117), Yanagida BioMotron Project, ERATO, JST, Senba-higasi 2-4-14, Mino, Osaka, 562 Japan

Michio Tomishige (173), Department of Life Sciences, Graduate School of Arts and Sciences, The University of Tokyo, Komaba 3-8-1, Meguro-ku, Tokyo 153, Japan

Bruce Tromberg (71), Beckman Laser Institute and Medical Clinic, University of California at Irvine, Irvine, California 92612

Watt W. Webb (99), School of Applied and Engineering Physics, Cornell University, Ithaca, New York 14853

Toshio Yanagida (117), Yanagida BioMotron Project, ERATO, JST, Senba-higasi 2-4-14, Mino, Osaka, 562 Japan, and Department of Physiology, Osaka University Medical School, Suita, Osaka, 565 Japan

[1] Present Address: Department of Physics and Astronomy, Swarthmore College, Swarthmore, Pennsylvania 19081

PREFACE

Over the past 10 years there has been an explosive growth in both the number of laser tweezers and the number of articles published with data using them. From sophisticated force measurements to simple placement of beads, there are a wide variety of applications for a device that enables the application of known forces to micrometer- and submicrometer-size particles even within cells. Dr. Arthur Ashkin's system for trapping bacteria was visionary and has opened the way for rapid measurements of force in a range difficult to measure otherwise. The purpose of this book is to provide the practical knowledge needed to set up a laser trap, to use it in a variety of applications, and to understand the trap limitations and potential pitfalls.

The practical range of forces that can be measured by the tweezers is from 0.2 to 200 pN for a particle approximately 1 μm in diameter. For larger forces, such as those needed to unfold most proteins, the atomic force microscope (AFM) is the current instrument of choice. Factors affecting the maximum force of the tweezers include the size of the particle (force increases approximately as the square of the particle diameter for 0.2- to 1.5-μm particles), the refractive index difference between the particle and the medium, the laser power (direct dependence), and the gradient of laser intensity in the microscope (any aberrations in beam profile will alter maximum force). In terms of understanding the theory behind the tweezers, the simplest access to the literature is through the Ashkin article (Chapter 1). For many motor and membrane experiments, the tweezers provide sufficient force and are more flexible than other system.

In terms of the practical issue of setting up laser tweezers, this book provides many useful hints, but nothing can truly replace hands-on experience. To get such experience, it is best to visit a laboratory currently working on applications similar to the one planned and see the chosen system in operation. Often a set of collaborative experiments will provide hands-on training as well as useful preliminary data. Like any other major tool, laser tweezers have limitations in each application. Before monetary and personnel resources are committed to acquiring laser tweezers, practical experience on a working system is often invaluable.

Although technology is developing at a rapid pace, the systems described in this book will be useful tools for solving many biological problems for years to come. Most cell functions have a mechanical component in either position dependence or force dependence that is difficult to probe without a tool like the laser tweezers. The one new area that will be important to watch is the overlap between laser tweezers and 2-photon confocal fluorescence. Trapped fluoresecent beads will fluoresce, and the exciting light for 2-photon fluorescence will exert considerable force on cellular structures. The area that may improve dramatically is lens design. Because many recent light microscope applications involve infrared wavelengths (800–1100 nm), new lenses may be designed to correct for chromatic aberrations at the end of the spectrum.

A variety of applications are considered here, but the list is by no means exhaustive. The trade-offs between the applications are considerable. If endogenous cellular structures or materials within cells are trapped and moved, the calibration of the force is difficult because neighboring structures and changes in the refractive index of cytoplasm can alter the applied force. Furthermore, the strength of the trap is insufficient to break cytoplasmic filaments (even one actin filament requires a force of 400 pN to break it). With extracellular beads or *in vitro* assays, the force per nanometer of bead displacement in the trap can be readily determined, but many cellular functions are difficult to approach outside the cell. Early cell biological applications of laser tweezers have been in the study of the plasma membrane because the forces on the particles can give information about critical cell functions directly. The other major current application of laser tweezers is in the area of biophysical, molecular mechanics, which includes motor function and polymer unfolding. As intracellular functions are recapitulated in *in vitro* assays, the forces involved can be readily measured using laser tweezers. Furthermore, laser tweezers can be used to spatially organize components *in vitro* to enable the reconstitution of cellular functions that require the correct juxtaposition of components.

In considering the current limitations of the tweezers from heating and photodamage to the time required for force analyses, we see possibilities for the greatest changes in the next 10 years in the area of analyses. Digital image analysis and quadrant detector improvements will essentially allow real-time analysis of forces even in complex samples. There are no obvious ways to overcome heating and photodamage limitations in cellular systems, but optimization of laser tweezers could conceivably give four- to eightfold increases in the maximum force per particle. Optimizing the wavelength of laser tweezers to minimize damage to the cells as well as optimizing the beam profile (such as the use of TEM-01 lasers) could enable greater forces to be applied for the same level of heating and photo damage. Thus, we can expect to be able to break single-actin filaments (400 pN) with laser tweezers, but it is unlikely that fibroblast focal contacts (>2 nN) will be broken. Still, a wide variety of cellular phenomena involve forces in the range of 2 to 400 pN, and rapid force analyses can allow us to define the molecular and physical bases of these phenomena.

Michael P. Sheetz

CHAPTER 1

Forces of a Single-Beam Gradient Laser Trap on a Dielectric Sphere in the Ray Optics Regime[1]

A. Ashkin
AT&T Bell Laboratories
Holmdel, New Jersey 07733

ABSTRACT We calculate the forces of single-beam gradient radiation pressure laser traps, also called "optical tweezers," on micron-sized dielectric spheres in the ray optics regime. This serves as a simple model system for describing laser trapping and manipulation of living cells and organelles within cells. The gradient and scattering forces are defined for beams of complex shape in the ray-optics limit. Forces are calculated over the entire cross-section of the sphere using TEM_{00} and TEM_{00}^{*} mode input intensity profiles and spheres of varying index of refraction. Strong uniform traps are possible with force variations less than a factor of 2 over the sphere cross-section. For a laser power of 10 mW and a relative index of refraction of 1.2, we compute trapping forces as high as $\sim 1.2 \times 10^{-6}$ dynes in the weakest (backward) direction of the gradient trap. It is shown that good trapping requires high convergence beams from a high numerical aperture objective. A comparison is given of traps made using bright field or differential interference contrast optics and phase contrast optics.

I. Introduction

This paper gives a detailed description of the trapping of micron-size dielectric spheres by a so-called single-beam gradient optical trap. Such dielectric spheres can serve as first simple models of living cells in biological trapping experiments and also as basic particles in physical trapping experiments. Optical trapping of small particles by the forces of laser radiation pressure has been used for about 20 yr in the physical sciences for the manipulation and study of micron and submicron dielectric particles and even individual atoms (*1–7*). These techniques have also been extended more recently to biological particles (*8–18*).

The basic forces of radiation pressure acting on dielectric particles and atoms are known (*1,2,19–21*). Dielectric spheres, large compared with the wavelength,

[1] This material may be protected by copyright law (Title 17 U.S. code). Reprinted with permission from A. Ashkin (1992). *Biophys. J.* **61,** 569–582.

METHODS IN CELL BIOLOGY, VOL. 55
Copyright © 1998 by Academic Press. All rights of reproduction in any form reserved.
0091-679X/98 $25.00

lie in the geometric optics regime; thus simple ray optics can be used in the derivation of the radiation pressure force from the scattering of incident light momentum. This approach was used to calculate the forces for the original trapping experiments on micron-size dielectric spheres (*1,22*). These early traps were all either optical two-beam traps (*1*) or single-beam levitation traps that required gravity or electrostatic forces for their stability (*23,24*). For particles in the Rayleigh regime in which the size is much less than the wavelength λ, the particle acts as a simple dipole. The force on a dipole divides itself naturally into two components: a so-called scattering force component pointing in the direction of the incident light and a gradient component pointing in the direction of the intensity gradient of the light (*19,21*).

The single-beam gradient trap, sometimes referred to as "optical tweezers," was originally designed for Rayleigh particles (*20*). It consists of a single strongly focused laser beam. Conceptually and practically it is one of the simplest laser traps. Its stability in the Rayleigh regime results from the dominance of the gradient force pulling particles toward the high focus of the beam over the scattering force trying to push particles away from the focus in the direction of the incident light. Subsequently it was found experimentally that single-beam gradient traps could also trap and manipulate micron-size (*25*) and a variety of biological particles, including living cells and organelles within living cells (*8,10*). Best results were obtained using infrared trapping beams to reduced optical damage. The trap in these biological applications was built into a standard high resolution microscope in which the same high numerical aperture (NA) microscope objective is used for both trapping and viewing. The micromanipulative abilities of single-beam gradient traps are finding use in a variety of experiments in the biological sciences. Experiments have been performed in the trapping of viruses and bacteria (*8*); the manipulation of yeast cells, blood cells, protozoa, and various algae and plant cells (*10*); the measurement of the compliance of bacterial flagella (*11*); internal cell surgery (*13*); manipulation of chromosomes (*12*); trapping and force measurement on sperm cells (*14,15*); and recently, observations on the force of motor molecules driving mitochondrion and latex spheres along microtubules (*16,17*). Optical techniques have also been used for cell sorting (*9*).

Qualitative descriptions of the operation of the single-beam gradient trap in the ray optics regimen have already been given (*25,26*). In Fig. 1 taken from reference *26*, the action of the trap on a dielectric sphere is described in terms of the total force due to a typical pair of rays a and b of the converging beam, under the simplifying assumption of zero surface reflection. In this approximation, the forces F_a and F_b are entirely due to refraction and are shown pointing in the direction of the momentum change. It can be seen that for arbitrary displacements of the sphere origin O from the focus f that the vector sum of F_a and F_b gives a net restoring force F directed back to the focus, and the trap is stable. In this paper we quantify the preceding qualitative picture of the trap. We show how to define the gradient and scattering force on a sphere $>> \lambda$ in a natural way

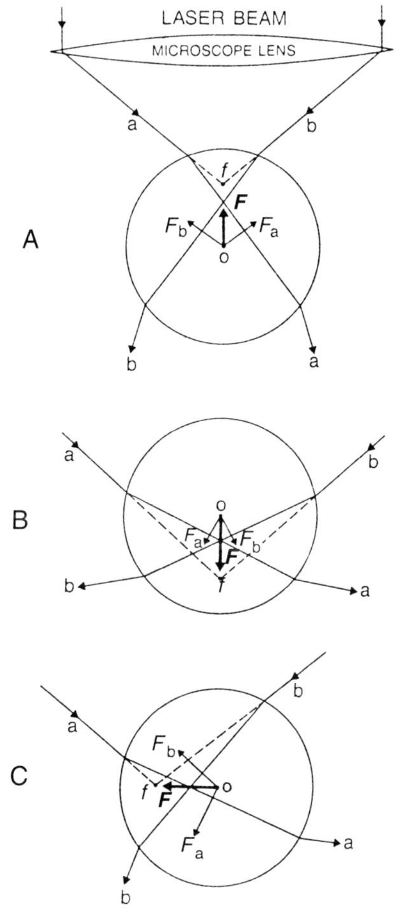

Fig. 1 Qualitative view of the trapping of dielectric spheres. The refraction of a typical pair of rays a and b of the trapping beam gives forces F_a and F_b whose vector sum F is always restoring for axial and transverse displacements of the sphere from the trap focus *f*.

for beams of arbitrary shape. Trapping in the ray optics regimen can then be described in the same terms as in the Rayleigh regimen. Results are given for the trapping forces over the entire cross-section of the sphere. The forces are calculated for input beams with various TEM_{00} and TEM_{01}^{*} mode intensity profiles at the input aperture of a high numerical aperture trapping objective (NA = 1.25). The results confirm the qualitative observation that good trapping requires the input aperture to be well enough filled by the incident beam to give rise to a trapping beam with a high convergence angle. Traps can be designed in which the trapping forces vary at most by a factor of ~1.8 over the cross-section of the sphere with trapping forces as high as $Q = 0.30$ where the force F is given in terms of the dimensionless factor Q in the expression $F = Q(n_1P/c)$. P is the incident power and n_1P/c is the incident momentum per second in a medium of index of refraction n_1. There has been a previous calculation of single-beam gradient trapping forces on spheres in the geometrical optics limit by Wright et al. (*27*) over a limited portion of the sphere, which gives much poorer results. These researchers found trapping forces of $Q = 0.055$ in the preceding units that vary over the sphere cross-section by more than an order of magnitude.

II. Light Forces in the Ray Optics Regime

In the ray optics or geometrical optics regime, the total light beam is decomposed into individual rays, each with appropriate intensity, direction, and state of polarization, which propagate in straight lines in media of uniform refractive index. Each ray has the characteristics of a plane wave of zero wavelength that can change directions when it reflects, refracts, and changes polarization at dielectric interfaces according to the usual Fresnel formulas. In this regimen diffractive effects are neglected (see Chapter 3 of reference 28).

The simple ray optics model of the single-beam gradient trap used here for calculating the trapping forces on a sphere of diameter $>> \lambda$ is illustrated in Fig. 2. The trap consists of an incident parallel beam of arbitrary mode structure and polarization that enters a high NA microscope objective and is focused ray-by-ray to a dimensionless focal point *f*. Fig. 2 shows the case in which *f* is located along the Z axis of the sphere. The maximum convergence angle for rays at the edge of the input aperture of a high NA objective lens such as the Leitz PL APO 1.25W (E. Leitz, Inc., Wetzlar, Germany) or the Zeiss PLAN NEOFLUAR 63/1.2W water immersion objectives (Carl Zeiss, Inc., Thornwood, NY), for example, is $\phi_{max} \cong 70°$. Computation of the total force on the sphere consists of summing the contributions of each beam ray entering the aperture at radius r with respect to the beam axis and angle β with respect the Y axis. The effect of neglecting the finite size of the actual beam focus, which can approach the limit of $\lambda/2n_1$ (see reference 29), is negligible for spheres much larger than λ. The point focus description of the convergent beam in which the ray directions and momentum continue in straight lines through the focus gives the correct incident polarization and momentum for each ray. The rays then reflect and refract at the surface of the sphere giving rise to the light forces.

The model of Wright et al. (*27*) rises to describe the single-beam gradient trap in terms of both wave and ray optics. It uses the TEM_{00} Gaussian mode beam propagation formula to describe the focused trapping beam and takes the directions of the individual rays to be perpendicular to the Gaussian beam phase fronts. Because the curvatures of the phase fronts vary considerably along the beam, the ray directions also change, from values as high as 30° or more with respect to the beam axis in the far field to 0° at the beam focus. This is physically incorrect. It implies that rays can change their direction in a uniform medium,

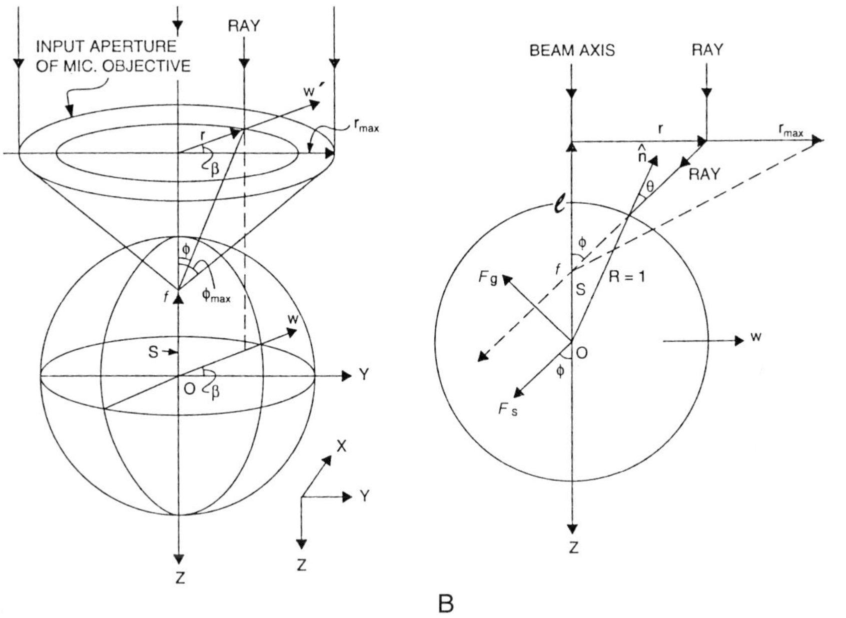

Fig. 2 (A) Single-beam gradient force trap in the ray optics model with beam focus f located along the Z axis of the sphere. (B) Geometry of an incident ray giving rise to gradient and scattering force contributions F_g and F_s.

which is contrary to geometrical optics. It also implies that the momentum of the beam can change in a uniform medium without interacting with a material object, which violates the conservation of light momentum. The constancy of the light momentum and ray direction for a Guassian beam can be seen in another way. If a Gaussian beam resolved into an equivalent angular distribution of plane waves (see Section 11.4.2 of reference *28*), it can be seen that these plane waves can propagate with no momentum or direction changes right through the focus. Another important point is that the Gaussian beam propagation formula is strictly correct only for transversely polarized beams in the limit of small far-field diffraction angles θ', where $\theta' = \lambda/\pi w_o$ (w_o being the focal spot radius). This formula therefore provides a poor description of the high convergence beams used in good traps. The proper wave description of a highly convergent beam is much more complex than the Gaussian beam formula. It involves strong axial electric field components at the focus (from the edge rays) and requires use of the vector wave equation as opposed to the scalar wave equation used for Gaussian beams (*30*).

Apart from the major differences near the focus, the model of Wright et al. (*27*) should be fairly close to the ray optics model used here in the far field of the trapping beam. The principal distinction between the two calculations, however, is the use by Wright et al. of beams with relatively small convergence angle. They calculate forces for beams with spot sizes $w_o = 0.5$, 0.6, and 0.7 μm, which implies values of θ' of ~29, 24, and 21°, respectively. Therefore, these beams have relatively small convergence angles compared with convergence angles of $\phi_{max} \cong 70°$, which are available from a high NA objective.

Consider first the force due to a single ray of power P hitting a dielectric sphere at an angle of incidence θ with incident momentum per second of n_1P/c (Fig. 3). The total force on the sphere is the sum of contributions due to the reflected ray of power PR and the infinite number of emergent refracted rays of successively decreasing power PT^2, PT^2R, . . . PT^2R^n, . . . The quantities R and T are the Fresnel reflection and transmission coefficients of the surface at θ. The net force acting through the origin O can be broken into F_Z and F_Y components as given by Roosen and co-workers (*3,22*) (see Appendix I for a sketch of the derivation):

$$F_Z = F_S = \frac{n_1P}{c} \left\{1 + R\cos 2\theta - \frac{T^2[\cos(2\theta - 2r) + R\cos 2\theta]}{1 + R^2 + 2R\cos 2r}\right\} \tag{1}$$

$$F_Y = F_g = \frac{n_1P}{c} \left\{R\sin 2\theta - \frac{T^2[\sin(2\theta - 2r) + R\sin 2\theta]}{1 + R^2 + 2R\cos 2r}\right\}, \tag{2}$$

where θ and r are the angles of incidence and refraction. These formulas sum over all scattered rays and are therefore exact. The forces are polarization dependent

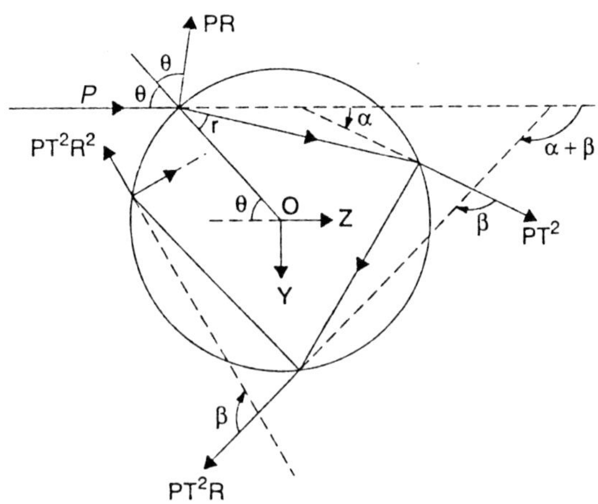

Fig. 3 Geometry for calculating the force due to the scattering of a single incident ray of power P by a dielectric sphere, showing the reflected ray PR and an infinite set of refracted rays PT^2R^n.

because R and T are different for rays polarized perpendicular or parallel to the plane of incidence.

In Eq. (1) we denote the E_z component pointing in the direction of the incident ray as the scattering force component F_s for this single ray. Similarly, in Eq. (2) we denote the F_y component pointing in the direction perpendicular to the ray as the gradient force component F_g for the ray. For beams of complex shape such as the highly convergent beams used in the single-beam gradient trap, we define the scattering and gradient forces of the beam as the vector sums of the scattering and gradient force contributions of the individual rays of the beam. Figure 2B depicts the direction of the scattering force component and gradient force component of a single ray of the convergent beam striking the sphere at angle θ. It can be shown that the gradient force, as defined, is conservative. This follows from the fact that F_g, the gradient force for a ray, can be expressed solely as a function of ρ, the radial distance from the ray to the particle. This implies that the integral of the work done on a particle in going around an arbitrary closed path can be expressed as an integral of $F_g(\rho)d\rho$, which is clearly zero. If the gradient force for a single ray is conservative, then the gradient force for an arbitrary collection of rays is conservative. Thus the conservative property of the gradient force as defined in the geometric optics regime is the same as in the Rayleigh regimen. The work done by the scattering force, however, is always path dependent and is not conservative in any regimen. As will be seen, these new definitions of gradient and scattering force for beams of more complex shape allow us to describe the operation of the gradient trap in the same manner in both the geometrical optics and Rayleigh regimens.

To get a feeling for the magnitudes of the forces, we calculate the scattering force F_s, the gradient force F_g, and the absolute magnitude of the total force $F_{mag} = (F_s^2 + F_g^2)^{1/2}$ as a function of the angle of incidence θ using Eqs. (1) and (2). We consider as a typical example the case of a circularly polarized ray hitting a sphere of effective index of refraction $n = 1.2$. The force for such a circularly polarized ray is the average of the forces for rays polarized perpendicular and parallel to the plane of incidence. The effective index of a particle is defined as the index of the particle n_2 divided by the index of the surrounding medium n_1; that is, $n = n_2/n_1$. A polystyrene sphere in water has $n = 1.6/1.33 \cong 1.2$. Figure 4 shows the results for the forces F_s, F_g, and F_{mag} versus θ expressed in terms of the dimensionless factors Q_s, Q_g, and $Q_{mag} = (Q_s^2 + Q_g^2)^{1/2}$, where

$$F = Q\frac{n_1 P}{c}. \qquad (3)$$

The quantity n_1P/c is the incident momentum per second of a ray of power P in a medium of index of refraction n_1 (*19,31*). Recall that the maximum radiation pressure force derivable from a ray of momentum per second n_1P/c corresponds to $Q = 2$ for the case of a ray reflected perpendicularly from a totally reflecting mirror. It can be seen that for $n = 1.2$ a maximum gradient force of Q_{gmax} as high as ~0.5 is generated for rays at angles of $\theta \cong 70°$. Table I shows the effect

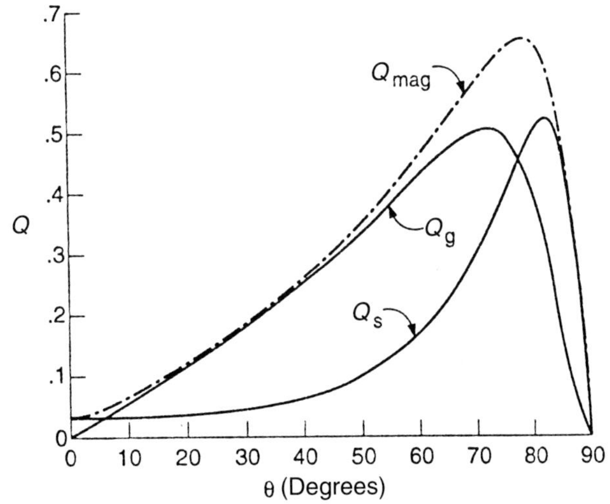

Fig. 4 Values of the scattering force Q_s, gradient force Q_g, and magnitude of the total force Q_{mag} for a single ray hitting a dielectric sphere of index of refraction $n = 1.2$ at an angle θ.

of an index of refraction n on the maximum value of gradient force Q_{max} occurring at an angle of incidence θ_{gmax}. The corresponding value of scattering force Q_s at θ_{gmax} is also listed. The fact that Q_s continues to grow relative to Q_{gmax} as n increases indicates potential difficulties in achieving good gradient traps at high n.

Table I
For a Single Ray. Effect of Index of Refraction n on Maximum Gradient Force Q_{gmax} and Scattering Force Q_s Occurring at Angle of Incidence θ_{gmax}

n	Q_{gmax}	Q_s	θ_{gmax}
1.1	−0.429	0.262	79°
1.2	−0.506	0.341	72°
1.4	−0.566	0.448	64°
1.6	−0.570	0.535	60°
1.8	−0.547	0.625	59°
2.0	−0.510	0.698	59°
2.5	−0.405	0.837	64°

III. Force of the Gradient Trap on Spheres

A. Trap Focus along Z Axis

Consider the computation of the force of a gradient trap on a sphere when the focus f of the trapping beam is located along the Z axis at a distance S above the center of the sphere at O, as shown in Fig. 2. The total force on the sphere, for an axially symmetric plane-polarized input trapping beam, is clearly independent of the direction of polarization by symmetry considerations. It can therefore be assumed for convenience that the input beam is circularly polarized with half the power in each of two orthogonally oriented polarization components. We find the force for a ray entering the input aperture of the microscope objective at an arbitrary radius r and angle β and then integrate numerically over the distribution of rays using an AT&T 1600 PLUS personal computer. As seen in Fig. 2, the vertical plane ZW, which is rotated by β from the ZY plane, contains both the incident ray and the normal to the sphere $\hat{n}$. It is thus the plane of incidence. We can compute the angle of incidence θ from the geometric relation $R \sin \theta = S \sin \phi$, where R is the radius of the sphere. We take $R = 1$ because the resultant forces in the geometric optics limit are independent of R. Knowing θ we can find F_g and F_s for the circularly polarized ray by first computing F_g and F_s for each of the two polarization components parallel and perpendicular to the plane of incidence using Eqs. (1) and (2) and adding the results. It is obvious by symmetry that the net force is axial. Thus for S above the origin O the contribution of each ray to the net force consists of a negative Z component $F_{gz} = -F_g \sin \phi$ and a positive Z component $F_{sz} = F_s \cos \phi$ as seen from Fig. 2B. For S below O the gradient force component changes sign and the scattering force component remains positive. We integrate out to a maximum radius r_{max} for which $\phi = \phi_{max} = 70°$, the maximum convergence angle for a water immersion objective of NA = 1.25, for example. Consider first the case of a sphere of index of refraction $n = 1.2$ and an input beam that uniformly fills the input aperture. Figure 5 shows the magnitude of the antisymmetric gradient force component, the symmetric scattering force component, and the total force, expressed as Q_g, Q_s, and Q_t, for values of S above and $(-S)$ below the center of the sphere. The sphere outline is shown in Fig. 5 for reference. It is seen that the trapping forces are largely confined within the spherical particle. The stable equilibrium point S_E of the trap is located just above the center of the sphere at $S \cong 0.06$, where the backward gradient force just balances the weak forward scattering force. Away from the equilibrium point the gradient force dominates over the scattering force, and Q_t reaches its maximum value very close to the sphere edges at $S \cong 1.01$ and $(-S) \cong 1.02$. The large values of net restoring force near the sphere edges are due to the significant fraction of all incident rays that have both large values of θ, near the optimum value of 70°, and large convergence angle θ. This assures a large backward gradient force contribution from the component F_g sin

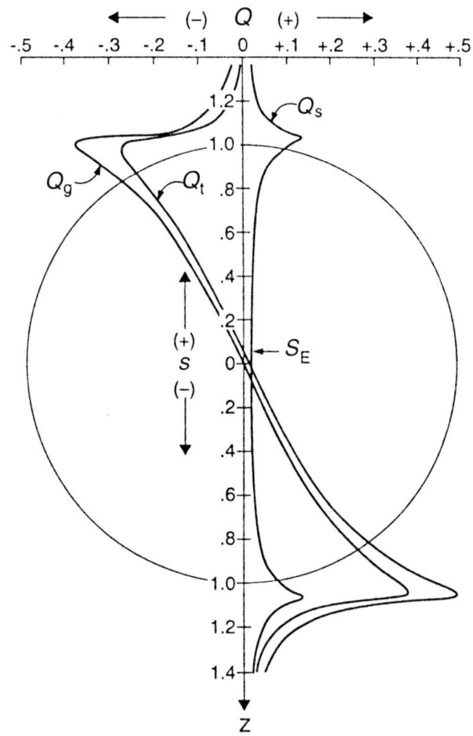

Fig. 5 Values of the scattering force, gradient force, and total force Q_s, Q_g, and Q_t exerted on a sphere of index of refraction $n = 1.2$ by a trap with a uniformly filled input aperture focused along the Z axis at positions $+s$ above and $-s$ below the center of the sphere.

ϕ and also a much-reduced scattering force contribution from the component $F_s \cos \phi$.

B. Trap along Y Axis

We next examine the trapping forces for the case where the focus f of the trapping beam is located transversely along the $-$Y axis of the sphere as shown in Fig. 6. The details of the force computation are discussed in Appendix II. Fig. 7 plots the gradient force, scattering force, and total force in terms of Q_g, Q_s, and Q_t as a function of the distance S' of the trap focus from the origin along the $-$Y axis for the same conditions as in III A. For this case the gradient force has only a $-$Y component. The scattering force is orthogonal to it along the +Z axis. The total force again maximizes at a value $Q_t \cong 0.31$ near the sphere edge at $S' \cong 0.98$ and makes a small angle $\phi = \arctan F_g/F_s \cong 18.5°$ with respect to the Y axis. The Y force is, of course, symmetric about the center of the sphere at O.

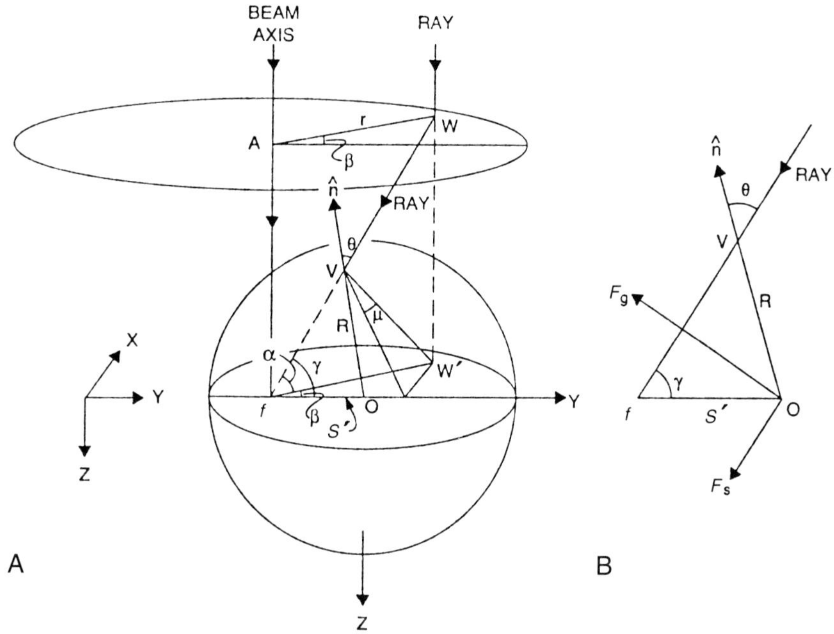

Fig. 6 (A) Trap geometry with the beam focus f located transversely along the −Y axis at a distance S' from the origin. (B) Geometry of the plane of incidence showing the directions of the gradient and scattering forces F_g and F_s for the input ray.

C. General Case: Arbitrary Trap Location

Consider finally the most general case in which the focus f is situated arbitrarily in the vertical plane through the Z axis at the distance S' from the sphere origin O in the direction of the −Y axis and a distance S'' in the direction of the −Z axis as shown in Fig. 8. Appendix III summarizes the method of force computation for this case.

Figure 10 shows the magnitude and direction of the gradient force Q_g, the scattering force Q_s, and the total force Q_t as the functions of the position of the focus f over the left half of the YZ plane, and by mirror image symmetry about the Y axis, over the entire cross-section of the sphere. This is again calculated for a circularly polarized beam uniformly filling the aperture and for $n = 1.2$. Although the force vectors are drawn at the point of focus f, it must be understood that the actual forces always act through the center of the sphere. This is true for all rays and therefore also for the full beam. It is an indication that no radiation pressure torques are possible on a sphere from the linear momentum of light. We see in Fig. 10A that the gradient force, which is exactly radial along the Z and Y axes, is also very closely radial (within an average of ~2° over the rest of the sphere. This stems from the closely radially uniform distribution of

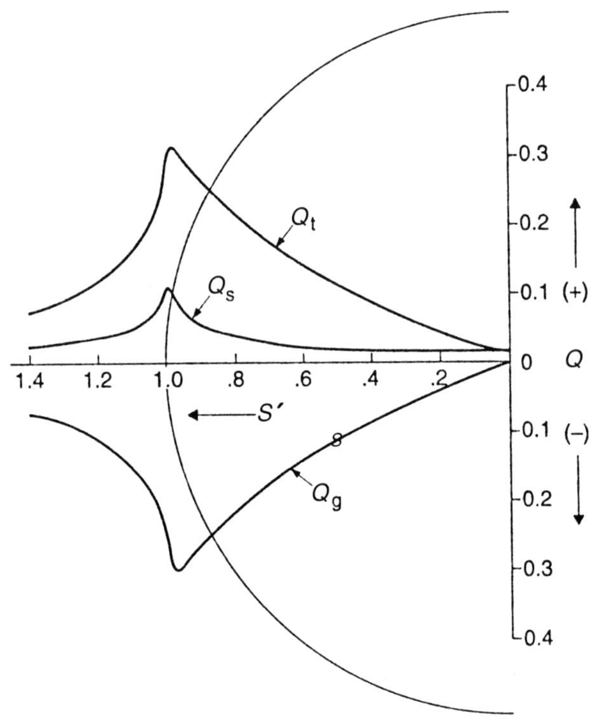

Fig. 7 Plot of the gradient force, scattering force, and total force Q_g, Q_s, and Q_t as a function of the distance S' of trap focus from the origin along the −Y axis or a circularly polarized trapping beam uniformly filling the aperture and a sphere of index of refraction $n = 1.2$.

the incident light in the upper hemisphere. The considerably smaller scattering force is shown in Fig. 10B (note the change in scale). It is strictly axial only along the Z and Y axes and remains predominantly axial elsewhere except for the regions farthest from the Z and Y axes. It is the dominance of the gradient force over the scattering force that accounts for the overall radial character of the total force in Fig. 10C. The rapid changes in direction of the force that occur when the focus is well outside the sphere are mostly due to the rapid changes in effective beam direction as parts of the input beam start to miss the sphere. We note that the magnitude of the total force Q_t maximizes very close to the edge of the sphere as we proceed radially outward in all directions, as does the

Fig. 8 (A) Trap geometry with the beam focus located at a distance S' from the origin in the −Y direction and a distance S'' in the −Z direction. (B) Geometry of the plane of incidence POV showing the direction of gradient and scattering forces F_g and F_s for the ray. Geometry of triangle POB in the XY plane for finding β' and d.

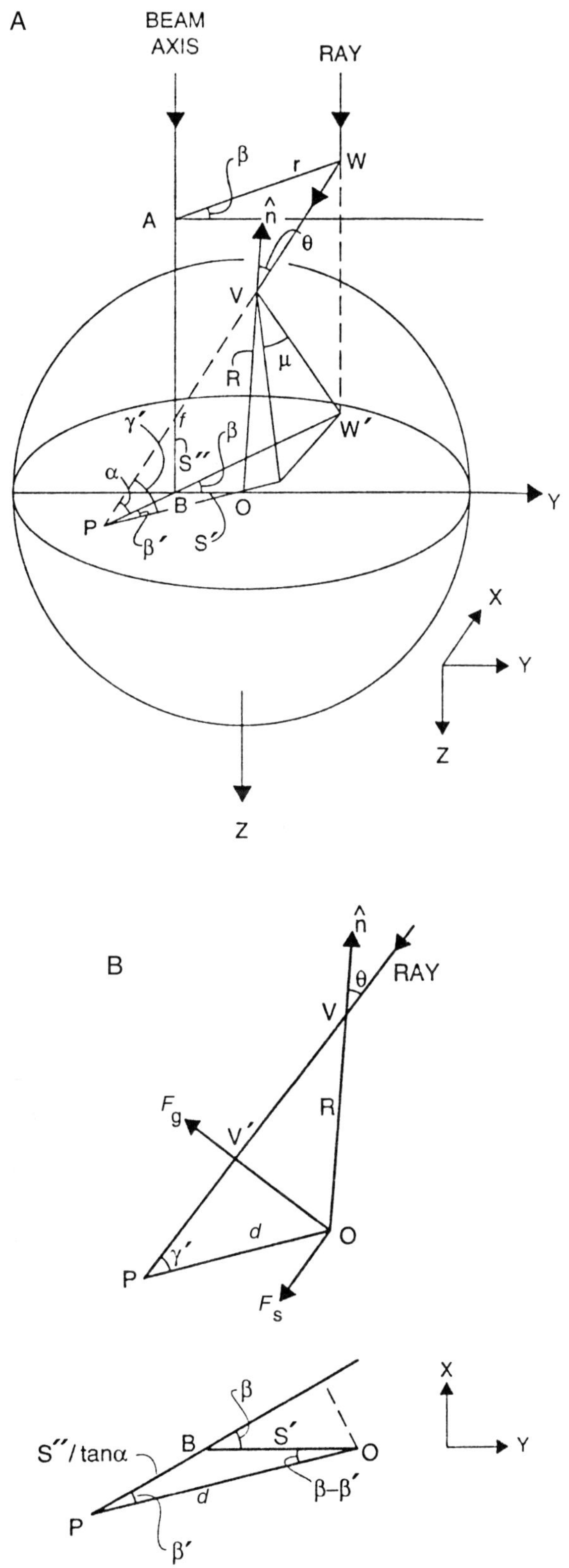
A
BEAM AXIS
RAY
β
r
W
A
n̂
θ
V
R
μ
γ′
f
S″
β
W′
α
B
O
Y
P
β′
S′
X
Y
Z
Z
B
n̂
θ
RAY
V
R
F_g
V′
d
O
γ′
P
F_s
β
S′
B
O
S″/tanα
β–β′
d
P
β′
X
Y

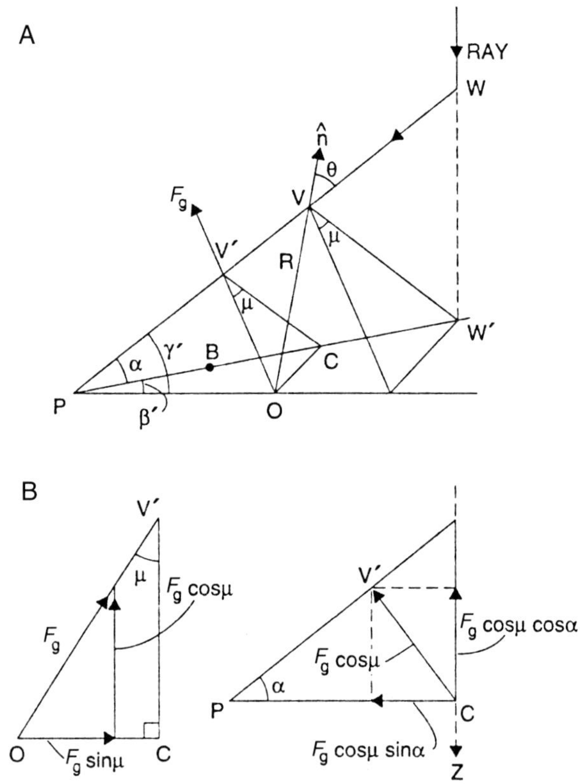

Fig. 9 Another view of Fig. 8 A containing the angle μ between the plane of incidence POV′ and the vertical plane WW′P for resolving force components along the coordinate axis.

gradient and scattering forces. The value of maximum restoring forces varies smoothly around the edge of the sphere from a maximum of $Q_t = 0.28$ in the axially backward direction to a maximum of $Q_t = 0.49$ in the forward direction. Thus, for these conditions the maximum trapping force achieved varies quite moderately over the sphere by a factor of $0.49/0.28 = 1.78$ and conforms closely to the edges of the sphere.

The line EE′ marked on Fig. 10C represents the locus of points for which the Z component of the force is zero (i.e., the net force is purely horizontal). If we start initially at point E, the equilibrium of the trap with no externally applied forces, and then apply a +Y-directed Stokes' force by flowing liquid past the sphere to the right, for example, the equilibrium position will shift to a new equilibrium point along EE′ where the horizontal light force just balances the viscous force. With increasing viscous force the focus finally moves to E′, the point of maximum transverse force, after which the sphere escapes the trap. Notice that there is a net z displacement of the sphere as the equilibrium point

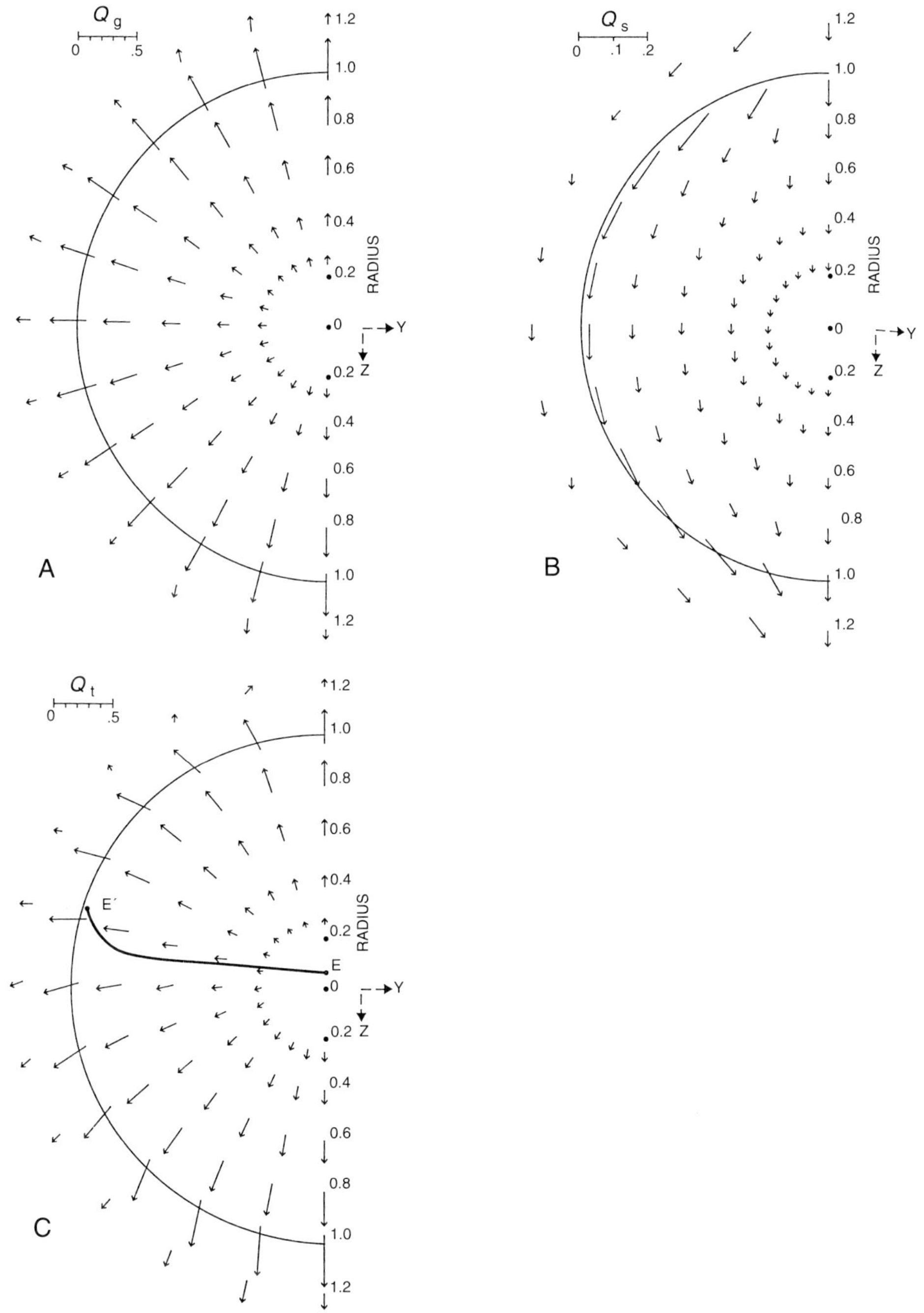

Fig. 10 A, B, and C show the magnitude and direction of gradient, scattering, and total force vectors Q_g, Q_s, and Q_t as a function of position of the focus over the YZ plane, for a circularly polarized trapping beam uniformly filling the aperture and a sphere of $n = 1.2$. Q_g is the vector sum of Q_g and Q_s. EE′ in C indicates the line along which Q_t is purely horizontal.

moves from E to E′. We have observed this effect in experiments with micron-size polystyrene spheres. Sato et al. (*18*) have recently reported also seeing this displacement.

IV. Effect of Mode Profiles and Index of Refraction on Trapping Forces

To achieve a uniformly filled aperture in practice requires an input TEM_{00}-mode Gaussian beam with very large spot size, which is wasteful of laser power. We therefore consider the behavior of the trap for other cases of TEM_{00}-mode input beam profiles with smaller spot sizes, as well as TEM_{01}^{*} "do-nut" mode beam profiles that preferentially concentrate input light intensity at large input angles ϕ.

A. TEM_{00}-Mode Profile

Table II compares the performance of traps with $n = 1.2$ having different TEM_{00}-mode intensity profiles of the form $I(r) = I_o \exp(-2r^2/w_o^2$ at the input aperture of the microscope objective. The quantity a is the ratio of the TEM_{00}-mode beam radius w_o to the full lens aperture r_{max}. A is the fraction of total beam power that enters the lens aperture. A decreases as a increases. In the limit of a uniform input intensity distribution, $A = 0$ and $a = \infty$. For $w_o \leq r_{max}$ we define the convergence angle of the input beam as θ' where $\tan\theta' = w_o/l$, in which l is the distance from the lens to the focus f as shown in Fig. 2B. For $w_o > r_{max}$ the convergence angle is set by the full lens aperture and we use $\theta' = \phi_{max}$, where $\tan\phi_{max} = r_{max}/l$. For a NA = 1.25, water immersion objective $\phi_{max} = 70°$. The quality of the trap can be characterized by the maximum strength of the restoring forces as we proceed radially outward for the sphere origin O in three representative directions taken along the Z and Y axes. We thus list Q_{1max}, the value of the maximum restoring force along the $-Z$ axis, and S_{max},

Table II
Performance of TEM_{30} Mode Tapes with $n = 1.2$ Having Different Intensity Profiles at the Input of the Microscopic Objective

a	A	$[Q_{1max}$,	$S_{max}]$	$[Q_{2max}$,	$S'_{max}]$	$[Q_{3max}$,	$(-S)_{max}]$	S_E	θ'
∞	0	−0.276	1.01	0.313	0.98	0.490	1.05	0.06	70°
1.7	0.5	−0.259	1.01	0.326	0.98	0.464	1.05	0.08	70°
1.0	0.87	−0.225	1.02	0.349	0.98	0.412	1.05	0.10	70°
0.727	0.98	−0.184	1.03	0.383	0.98	0.350	1.06	0.13	63°
0.364	1.0	−0.077	1.15	0.498	0.98	0.214	1.3	0.32	45°
0.202	1.0	−0.019	1.4	0.604	0.98	0.147	1.9	0.80	29°

the radial distance from the origin at which it occurs. Similarly listed are Q_{2max} occurring at S'_{max} along the $-$Y axis and Q_{3max} occurring at $(-S)_{max}$ along the +Z axis (see Figs. 2, 5, and 6 for a reminder on the definitions of S, $-S$, and S'). S_E in Table II gives the location of the equilibrium point of the trap along the $-$Z axis as noted in Fig. 5.

It can be seen from Table II that the weakest of the three representative maximum restoring forces is Q_{1max} occurring in the $-$Z direction. Furthermore, of all the traps the $a = \infty$ trap with a uniformly filled aperture has the largest Q_{1max} force and is therefore the strongest of all the TEM_{00}-mode traps. The "escape force" of a given trap can be defined as the lowest force that can pull the particle free of the trap in any direction. In this context the $a = \infty$ trap has the largest magnitude of escape force of $Q_{1max} = 0.276$. It also can be seen that the $\alpha = \infty$ trap is the most uniform trap because it has the smallest fractional variation in the extreme values of the restoring forces Q_{1max} and Q_{3max}. If, however, we reduce a to 1.7 or even 1.0, where the fraction of input power entering the aperture is reasonably high ($\sim$ 0.50 or 0.87), we can still get performance close to that of the uniformly filled aperture. Trap performance, however, rapidly degrades for cases of underfilled input aperture and decreasing beam convergence angle. For example, in the trap with $a = 0.202$ and $\theta' \cong 29°$ the value of Q_{1max} has dropped more than an order of magnitude to $Q_{1max} = -0.019$. The maximum restoring forces Q_{1max} and Q_{2max} occur well outside the sphere, and the equilibrium position has moved away from the origin to $S_E = 0.8$. This trap with $\theta' \cong 29°$ roughly corresponds to the best of the traps described by Wright et al. (*27*) (for the case of $w_o = 0.5$ μm). These researchers found for $w_o = 0.5$ μm that the trap has an equilibrium position outside of the sphere and a maximum trapping force equivalent to $Q_{1max} = -0.055$. Any more direct comparison of our results with those of Wright et al. is not possible since they use an approximate force calculation that overestimates the forces somewhat. They do not calculate forces for the beam focus inside the sphere, and there are other artifacts associated with their use of Gaussian beam phase fronts to give the incident ray directions near the beam focus.

B. TEM^{*}_{01} Do-nut-Mode Profile

Table III compares the performance of several traps based on the TEM^{*}_{01} mode, the so-called "do-nut" mode, which has an intensity distribution of the form $I(r) = I_o\,(r/w'_o)^2 \exp(-2r^2/w'_o)^2$. The quantity a is now the ratio of w'_o, the spot size of the do-nut mode, to the full lens aperture r_{max}. All other items in the Table III are the same as these in Table II. For $a = 0.76$, ~87% of the total beam power enters the input aperture r_{max} and we obtain performance that is almost identical to that of the trap with uniformly filled aperture as listed in Table II. For larger values of a the absolute magnitude of Q_{1max} increases, the magnitude of Q_{2max} decreases, and the fraction of power entering the aperture decreases. Optimal trapping, corresponding to the highest value of escape force,

Table III
Performance of TEM$^{*}_{01}$ Mode Traps with n = 1.2 Having Different Intensity Profiles at the Input of the Microscope Objective

a	A	$[Q_{max}$,	$S_{max}]$	$[Q_{max}$,	$S'_{max}]$	$[Q_{max}$,	$(-S)_{max}]$	S_E
			TEM$^{*}_{01}$ do-nut mode traps					
1.21	0.40	−0.310	1.0	0.290	0.98	0.544	1.05	0.06
1.0	0.59	−0.300	1.01	0.296	0.98	0.531	1.05	0.06
0.938	0.66	−0.296	1.01	0.298	0.98	0.525	1.05	0.07
0.756	0.87	−0.275	1.01	0.311	0.98	0.494	1.06	0.10
			Ring beam with ϕ = 70°					
		−0.366	0.99	0.254	0.95	0.601	1.03	
			Ring beam plus axial beam					
		−0.31	0.99	0.31	0.95	0.51	1.03	

Comparison data on a ring beam having ϕ = 70° and a ring beam plus an axial beam containing 18% of the power.

is achieved at values of $a \cong 1.0$ where the magnitudes $Q_{1max} \cong Q_{2max} \cong 0.30$. This performance is somewhat better than that achieved with TEM$_{00}$-mode traps.

It is informative to compare the performance of do-nut mode traps with that of a so-called "ring trap," which has all its power concentrated in a ring 95 to 100% of the full beam aperture, for which $\phi \cong \phi_{max} = 70°$. When the ring trap is focused at $S \cong 1.0$, essentially all of the rays hit the sphere at an angle of incidence very close to $\theta_{gmax} = 72°$, the angle that makes Q_g a maximum for $n = 1.2$ (see Table I). Thus the resulting backward total force of $Q_{1max} = 0.366$ at $S = 0.99$, as listed in Table III, closely represents the highest possible backward force on a sphere of $n = 1.2$. The ring trap, however, has a reduced force $Q_{2max} = 0.254$ at $S'_{max} = 0.95$ in the −Y direction because many rays at this point are far from optimal. If we imagine adding an axial beam to the ring beam, then we optimally increase the gradient contribution to the force in the −Y direction near $S' = 1.0$ and decrease the overall force in the −Z direction. With 18% of the power in the axial beam we get $Q_{1max} = Q_{2max} \cong 0.31$. This performance is now close to that of the optimal do-nut mode trap. It is possible to design gradient traps that approximate the performance of a ring trap using a finite number of individual beams (e.g., four, three, or two beams) located symmetrically about the circumference of the ring and converging to a common focal point at angles of $\phi \cong 70°$. Recent reports (*32,33*) at the CLEO-'91 conference presented observations on a trap with two individual beams converging to a focus with $\phi \cong 65°$ and also on a single beam gradient trap using the TEM$^{*}_{01}$ mode.

Knowledge of the forces produced by ring beams allows comparison of the forces generated by bright-field microscope objectives, as have thus far been considered, with the forces from phase-contrast objectives of the same NA. For example, assume a phase contrast objective having an 80% absorbing phase ring located between radii of 0.35 and 0.55 of the full input lens aperture. For the case of an input beam uniformly filling the aperture with $n = 1.2$, we find that the bright-field escape force of $Q_{1max} = 0.276$ (see Table II) increases by ~4% to $Q_{1max} = 0.287$ in going to the phase-contrast objective. With a TEM_{00}-mode Gaussian beam input having $A = 0.87$ and $n = 1.2$, the bright-field escape force magnitude of $Q_{1max} = 0.225$ increases by ~2% to $Q_{1max} = 0.230$ for a phase-contrast objective. The reason for these slight improvements is that the force contribution of rays at the ring corresponds to $Q_{1max} \cong 0.204$, which is less than the average force for bright field. Thus any removal of power at the ring radius improves the overall force per unit transmitted power. Differential interference contrast optics can make use of the full input lens aperture and thus gives equivalent trapping forces to bright-field optics.

C. Index of Refraction Effects

Consider, finally, the role of the effective index of refraction of the particle $n = n_1/n_2$ on the forces of a single-beam gradient trap. In Table IV we vary n for two types of trap, one with a uniformly filled input aperture, and the other having a do-nut input beam with $a = 1.0$, for which the fraction of total power feeding the input aperture is 59%. For the case of the uniformly filled aperture we get good performance over the range $n = 1.05$ to $n \cong 1.5$, which covers the regimen of interest for most biological samples. At higher index, Q_{1max} falls to a value of 0.097 at $n = 2$. This poorer performance is due to the increasing scattering force relative to the maximum gradient force as n increases (see Table

Table IV
Effect of Index Refraction *n* on the Performance of a Trap with a Uniformly Filled Aperture ($a = \infty$) and a Do-nut Trap with $a = 1.0$

n	[Q_{1max},	S_{max}]	[Q_{2max},	S'_{max}]	[Q_{3max},	$(-S)_{max}$]	S_E
			Trap with uniformly filled aperture				
1.05	−0.171	1.06	0.137	1.00	0.219	1.06	0.02
1.1	−0.231	1.05	0.221	0.99	0.347	1.06	0.04
1.2	−0.276	1.01	0.313	0.98	0.490	1.05	0.06
1.3	−0.288	0.96	0.368	0.97	0.573	1.04	0.11
1.4	−0.282	0.93	0.403	0.96	0.628	1.02	0.15
1.6	−0.237	0.89	0.443	0.94	0.693	1.00	0.25
1.8	−0.171	0.88	0.461	0.94	0.723	0.99	0.37
2.0	−0.097	0.88	0.469	0.94	0.733	0.99	0.53
			TEM^*_{01} do-nut mode trap with $a = 1.0$				
1.05	−0.185	1.06	0.134	1.00	0.238	1.06	0.02
1.1	−0.250	1.05	0.208	0.99	0.379	1.06	0.03
1.2	−0.300	1.01	0.296	0.98	0.531	1.05	0.06
1.4	−0.309	0.93	0.382	0.95	0.667	1.02	0.13
1.8	−0.204	0.88	0.434	0.94	0.748	0.99	0.32
2.0	−0.132	0.88	0.439	0.94	0.752	0.99	0.42

I). Also, the angle of incidence for maximum gradient force falls for higher n. At $n = 2$ (which corresponds roughly to a particle of index ~2.7 in water of index 1.33), the do-nut mode trap is clearly better than the uniform beam trap.

V. Concluding Remarks

We have shown how to define the gradient and scattering forces acting on dielectric spheres in the ray optics regime for beams of complex shape. The operation of single-beam gradient-force traps can then be described for spheres of diameter ≫ in terms of the dominance of an essentially radial gradient force over the predominantly axial scattering force. This is analogous to the previous description of the operation of this trap in the Rayleigh regimen, where the diameter ≪. Quite strong uniform traps are possible for $n = 1.2$ using the TEM_{01}^{*} do-nut mode in which the trapping forces vary over the sphere cross-section from a Q value of −0.30 in the −Z direction to 0.53 in the +Z direction. The magnitude of trapping force of 0.30 in the weakest trapping direction gives the escape force which a spherically shaped motile living organism, for example, must exert to escape the trap. For a laser power of 10 mW the minimum trapping force or escape force of $Q = 0.30$ is equivalent to 1.2×10^{-6} dynes. This implies that a motile organism 10 μm in diameter, which is capable of propelling itself through water at a speed of 128 μm/sec, will be just able to escape the trap in its weakest direction along the −Z axis. The only possible drawback to using the do-nut mode in practice is the difficulty of generating that mode in the laser. With the simpler TEM_{00} mode beams traps with Q's as high as 0.23 can be achieved, for example, with 87% of the laser power entering the aperture of the microscope objective.

The calculation confirms the importance of using beams with large convergence angles θ' as high as ~70° for achieving strong traps, especially with particles having lower indices of refraction typical of biological samples. At small convergence angles, less than ~30°, the scattering force dominates over the gradient force and single-beam trapping is either marginal or not possible. However, a two-beam gradient-force trap can be made using smaller convergence angles based on two confocal, oppositely directed beams of equal power in which each ray of the converging beam is exactly matched by an oppositely directed ray. Then the scattering forces cancel and the gradient forces add, giving quite a good trap. Gradient traps of this type have been previously observed in experiments on alternating-beam traps (*34*). The advantage of lower beam convergence is the ability to use longer working distances.

This work using ray optics extends the quantitative description of the single-beam gradient trap for spheres to the size regime in which the diameter is ≫. In this regime the force is independent of particle radius r. In the Rayleigh regimen the force varies as r^3. At present there is no quantitative calculation for the intermediate size regime in which the diameter is $\approx \lambda$, for which we expect force variations between r^0 and r^3. This is a more difficult scattering problem and involves an extension of Mie theory (*35*) or vector methods (*36*) to the case of highly convergent beams. Experimentally, however, this intermediate regime presents no problems. We can often directly calibrate the magnitude of the trapping force using Stokes' dragging forces and thus successfully perform experiments with biological particles of size $\approx\lambda$ (*16*).

We can get a good idea of the range of validity of the trapping forces as computed in the ray optics regimen from a comparison of the scattering of a plane wave by a large dielectric sphere in the ray optics regimen with the exact scattering, including all diffraction effects, as given by Mie theory. It suffices to consider plane waves because complex beams can be decomposed into a sum of plane waves. It was shown by van de Hulst in Chapter 12 of his book (*35*) that ray optics give a reasonable approximation to the exact angular intensity distribution of Mie theory (except in a few special directions) for sphere size parameters $2\pi r/\lambda = 10$ or 20. The special directions are the forward direction, in which a large diffraction peak appears that contributes nothing to the radiation pressure, and the so-called glory and rainbow directions, in which ray optics never works. Because these directions contribute only slightly to the total force, we expect ray optics to give fair results down to diameters of approximately six wavelengths or ~5 μm for a 1.06-μm laser beam in water. The validity of the approximation should improve rapidly at larger sphere diameters. A similar result was also derived by van de Hulst (*35*) using Fresnel zones to estimate diffractive effects.

One advantage of a reliable theoretical value for the trapping force is that it can serve as a reference for comparison with experiment. If discrepancies appear in such a comparison, we can then look for the presence of other forces. For traps using infrared beams, there could be significant thermal (radiometric) force contributions due to absorptive heating of the particle or surrounding medium, whose magnitude could then be inferred. Detailed knowledge of the variation of trapping force positioned within the sphere is also proving useful in measurements of the force of swimming sperm (*15*).

Appendix I: Force of a Ray on a Dielectric Sphere

A ray of power P hits a sphere at an angle θ where it partially reflects and partially refracts, giving rise to a series of scattered rays of power PR, PT^2, PT^2R, . . . , PT^2R^n, . . . As seen in Fig. 3, these scattered rays make angles relative to the incident forward ray direction of $\pi + 2\phi$, α, $\alpha + \beta$, . . . , $\alpha + \eta\beta$ . . . , respectively. The total force in the Z direction is the net change in momentum per second in the Z direction due to the scattered rays. Thus

$$F_z = \frac{n_1 P}{c} - \left[\frac{n_1 PR}{c} \cos(\pi + 2\theta) + \sum_{n=0}^{\infty} \frac{n_1 P}{c} T^2 R^n \cos(\alpha + n\beta) \right], \qquad \text{(A1)}$$

where $n_1 P/c$ is incident momentum per second in the Z direction. Similarly for the Y direction, where the incident momentum per second is zero, one has

$$F_Y = 0 - \left[\frac{n_1 PR}{c} \sin(\pi + 2\theta) - \sum_{n=0}^{\infty} \frac{n_1 P}{c} T^2 R^n \sin(\alpha + \beta) \right]. \qquad \text{(A2)}$$

As pointed out by van de Hulst in Chapter 12 of reference 35 and by Roosen (*22*), the rays scattered by a sphere can be summed over by considering the total force in the complex plane, $F_{tot} = F_z + iF_Y$. Thus

$$F_{tot} = \frac{n_1 P}{c}[1 + R \cos 2\theta] + i\frac{n_1 P}{c} R \sin 2\theta - \frac{n_1 P}{c} T^2 \sum_{n=0}^{\infty} R^n e^{i(\alpha + n\beta)}. \qquad \text{(A3)}$$

The sum over n is a simple geometric series that can be summed to give

$$F_{tot} = \frac{n_1 P}{c}[1 + R \cos 2\theta] + i\frac{n_1 P}{c} R \sin 2\theta - \frac{n_1 P}{c} T^2 e^{i\alpha} \left[\frac{1}{1 - Re^{i\beta}} \right]. \qquad \text{(A4)}$$

If we rationalize the complex denominator and take the real and imaginary parts of F_{tot}, we get the force expressions A1 and A2 for F_z and F_Y using the geometric relations $\alpha = 20 - 2r$ and $\beta = \pi - 2r$, where θ and r are the angles of incidence and refraction of the ray.

Appendix II: Force on a Sphere for Trap Focus along Y Axis

We treat the case of the beam focus located along the $-$Y axis at a distance S' from the origin O (see Fig. 6). We first calculate the angle of incidence θ for an arbitrary ray entering the input lens aperture vertically at a radius r and azimuthal angle β in the first quadrant. On leaving the lens the ray stays in the vertical plane AWW′ f and heads in the direction toward f, striking the sphere at V. The forward projection of the ray makes an angle α with respect to the horizontal (X, Y) plane. The plane of incidence, containing both the input ray and that normal to the sphere OV, is the so-called γ plane fOV that meets the horizontal and vertical planes at f. Knowing α and β, we find γ from the geometrical relation $\cos \gamma = \cos \alpha \cos \beta$. Referring to the γ plane we can now find the angle of incidence θ from $R \sin \theta = S' \sin \gamma$ putting $R = 1$.

In contrast to the focus along the Z axis, the net force now depends on the choice of input polarization. For the case of an incident beam polarized perpendicular to the Y axis, for example, the polarized electric field E is first resolved into components E $\cos \beta$ and E $\sin \beta$ perpendicular and parallel to the vertical plane containing the ray. Each of these components can be further resolved into the so-called p and s components parallel and perpendicular to the plane of incidence in terms of angle μ between the vertical plane and the plane of incidence. By geometry, $\cos \mu = \tan \alpha / \tan \gamma$. This resolution yields fractions of the input power in the p and s components given by

$$f_p = (\cos \beta \sin \mu - \sin \beta \cos \mu)^2, \quad \text{(A5)}$$
$$f_s = (\cos \beta \cos \mu + \sin \beta \sin \mu)^2. \quad \text{(A6)}$$

If the incident polarization is parallel to the Y axis, then f_p and f_s reverse. If θ, f_p, and f_s, are known, the gradient and scattering force components for p and s are computed separately using Eqs. (A5) and (A6), and the results are added.

The net gradient and scattering force contribution of the ray thus computed must now be resolved into components along the coordinate axes (see Fig. 6B). However, comparing the force contributions of the quartet of rays made up of the ray in the first quadrant and its mirror image rays in the other quadrants we see that the magnitudes of the forces are identical for each ray of the quartet. Furthermore, the scattering and gradient forces of the quartet are directly symmetrically about the Z and Y axes, respectively. This symmetry implies that the entire beam can give rise only to a net Z scattering force coming from the integral of the $F_s \cos \phi$ component and a net Y gradient force coming from the $F_g \sin \gamma$ component. In practice we need only integrate these components over the first quadrant and multiply the results by 4 to get the net force. The differences in force that result from the choice of input polarization perpendicular or parallel to the Y axis are not large. For the conditions of Fig. 7 the maximum force difference is ~14% near $S' \cong 1.0$. We have therefore made calculations using a circularly polarized input beam with $f_p = f_s = \frac{1}{2}$, which yields values of net force that are close to the average of the forces for the two orthogonally polarized beams.

Appendix III: Force on a Sphere for an Arbitrarily Located Trap Focus

We now treat the case in which the trapping beam is focused arbitrarily in the XY plane at a point f located at a distance S' from the origin in the $-$Y direction and a distance S'' in the $-$Z direction (see Fig. 8). To calculate the force for a given ray we again need to find the angle of incidence θ and the fraction of the ray's power incident on the sphere in the s and p polarizations. Consider a ray of the incident beam entering the input aperture of the lens vertically at a radius r and azimuthal angle β in the first quadrant. The ray on leaving the lens stays in the vertical plane AWW'B and heads toward f, hitting the sphere at V. The extension of the incident ray to f and beyond intersects the XY plane at point P at an angle α. The plane of incidence for this ray is the so-called γ' plane POV, which contains both the incident ray and that normal to the sphere OV. Referring to the planar figure in Fig. 8B can one find the angle β' by simple geometry in terms of S', S'', and the known angles α and β from the relation

$$\tan \beta' = \frac{S' \sin \beta}{S' \cos \beta + S''/\tan \alpha}. \tag{A7}$$

We get γ' from $\cos \gamma' = \cos \alpha \cos \beta'$. Referring to the γ' plane in Fig. 8B we get the angle of incidence θ for the ray from $R \sin \theta = d \sin \gamma'$, putting $R = 1$. The distance d is deduced from the geometric relation

$$d = \frac{S'' \cos \beta'}{\tan \alpha} + S' \cos(\beta - \beta'). \tag{A8}$$

As in Appendix II, we compute f_p and f_s, the fraction of the ray's power in the p and s polarizations, in terms of the angle μ between the vertical plane W′VP and the plane of incidence POV. We use Eqs. (A5) and (A6) for the case of a ray polarized perpendicular to the Y axis and the same expressions with f_p and f_s reversed for a ray polarized parallel to the Y axis. To find μ we use $\cos \mu = \tan \alpha/\tan \gamma'$. As in Appendix II we can put $f_p = f_s = \frac{1}{2}$ and get the force for a circularly polarized ray, which is the average of the force for the cases of two orthogonally polarized rays.

The geometry for resolving the net gradient and scattering force contribution of each ray of the beam into components along the axes is now more complex. The scattering force F_s is directed parallel to the incident ray in the VP direction of Fig. 8. It has components $F_s \sin \alpha$ in the +Z diretion and $F_s \cos \alpha$ pointing in the BP direction in the XY plane. $F_s \cos \alpha$ is then resolved with the help of Fig. 8B into $F_s \cos \alpha \cos \beta$ in the −Y direction and $F_s \cos \alpha \sin \beta$ in the −X direction. The gradient force F_g points in the direction OV′ perpendicular to the incident ray direction VP in the plane of incidence OPV. This is shown in Fig. 8 and also in Fig. 9, which gives yet another view of the geometry. In Fig. 9 we consider the plane V′OC, which is taken perpendicularly to the γ' plane POV and the vertical plane WW′P. This defines the angle OV′C as μ, the angle between the planes, and also makes the angles OCV′, OCP, and CV′P right angles. As an aid to visualization we can construct a true three-dimensional model out of cardboard of the geometric figure for the general case as shown in Figs. 8 and 9. Such a model will make it easy to verify that the aforementioned angles are indeed right angles, and to see other details of the geometry. We can now resolve F_g into components along the X, Y, and Z axes with the help of right triangles OV′C and CV′P as shown in Fig. 9B. In summary, the net contribution of a ray in the first quadrant to the force is

$$F(\mathrm{Z}) = F_s \sin \alpha + F_g \cos \mu \cos \alpha \tag{A9}$$

$$F(\mathrm{Y}) = -F_s \cos \alpha \cos \beta + F_g \cos \mu \sin \alpha \cos \beta + F_g \sin \mu \sin \beta \tag{A10}$$

$$F(\mathrm{X}) = -F_s \cos \alpha \sin \beta + F_g \cos \mu \sin \alpha \sin \beta - F_g \sin \mu \cos \beta. \tag{A11}$$

The force Eqs. (A9–A11) are seen to have the correct signs because F_s and F_g are, respectively, positive and negative as calculated from Eqs. (1) and (2).

For the general case under consideration we lose all symmetry between first and second quadrant forces, and we must extend the force integrals into the second quadrant. All the preceding formulas derived for rays of the first quadrant are equally correct in the second quadrant using the appropriate values of the angles β, β', γ', and μ. For example, in the second quadrant β' can be obtuse. This gives obtuse γ' and obtuse μ. Obtuse μ implies that the γ' plane has rotated its position beyond the perpendicular to the vertical plane AWW'. In this orientation the gradient force direction tips below the XY plane and reverses its Z component as indicated by the sign change in the $F_g \cos \mu \cos \alpha$ term.

There are, however, some symmetry relations in the force contributions of rays of the input beam that still apply. For example, there is symmetry about the Y axis, that is, rays of the third and fourth quadrants give the same contribution to the Z and Y forces as rays of the first and second quadrants, whereas their X contributions exactly cancel. To find the net force we need only integrate the Y and Z components of first and second quadrants and double the result.

If we make S'' negative in all formulas, we obtain the correct magnitudes and directions of the forces for the case of the focus below the XY plane. Although we find different total force values for S'' positive and S'' negative, (i.e., symmetrical beam focus points above and below the XY plane), there still are symmetry relations that apply to the scattering and gradient forces separately. Thus we find that the Z components of the scattering force are the same above and below, but the Y component reverses. For the gradient force the Z components reverse above and below, and the Y components are the same. This is seen to be true in Fig. 10. It is also consistent with Fig. 5 showing the forces along the Z axis. This type of symmetrical behavior arises from the fact that the angle of incidence for rays entering the first quadrant from above the XY plane (S'' positive) is the same as for symmetrical rays entering in the second quadrant below the XY plane (S'' negative). Likewise the angles of incidence are the same for the second quadrant above and the first quadrant below. These results permit the force below the XY plane to be directly deduced from the values computed above the XY plane. The results derived here for the focus placed at an arbitrary point within the YZ plane are perfectly general because we can always choose to calculate the force in the cross-sectional plane through the Z axis that contains the focus f.

As a check on the calculations we can show that the results putting $S'' = 0$ in the general case are identical with those from the simpler Y axis integrals derived in Appendix II. Also in the limit $S' \rightarrow 0$ one gets the same results as those given by the simpler Z axis integral discussed earlier.

References

1. Ashkin, A. (1970). Acceleration and trapping of particles by radiation pressure. *Phys. Rev. Lett.* **24,** 156–159.

2. Ashkin, A. (1970). Atomic-beam deflection by resonance-radiation pressure. *Phys. Rev. Lett.* **24,** 1321–1324.
3. Roosen, G. (1979). Optical levitation of spheres. *Can. J. Phys.* **57,** 1260–1279.
4. Ashkin, A. (1980). Applications of laser radiation pressure. *Science* (*Wash, DC*) **210,** 1081–1088.
5. Chu, S., Bjorkholm, J. E., Ashkin, A., and Cable, A. (1986). Experimental observation of optically trapped atoms. *Phys. Rev. Lett.* **57,** 314–317.
6. Chu, S., and Wieman, C. (1989). Feature editors, special edition, laser cooling and trapping of atoms. *J. Opt. Soc. Am.* **B6,** 2020–2278.
7. Misawa, H., Koshioka, M., Sasaki, K., Kitamura, N., and Masuhara, H. (1990). Laser trapping, spectroscopy, and ablation of a single latex particle in water. *Chem. Lett.* **8,** 1479–1482.
8. Ashkin, A., and Dziedzic, J. M. (1987). Optical trapping and manipulation of viruses and bacteria. *Science* (*Wash. DC*) **235,** 1517–1520.
9. Buican, T., Smith, M. J., Crissman, H. A., Salzman, G. C., Stewart, C. C., and Martin, J. C. (1987). Automated single-cell manipulation and sorting by light trapping. *Appl. Opt.* **26,** 5311–5316.
10. Ashkin, A., Dziedzic, J. M., and Yamane, T. (1987). Optical trapping and manipulation of single cells using infrared laser beams. *Nature* (*Lond.*). **330,** 769–771.
11. Block, S. M., Blair, D. F., and Berg, H. C. (1989). Compliance of bacterial flagella measured with optical tweezers. *Nature* (*Lond.*). **338,** 514–518.
12. Berns, M. W., Wright, W. H., Tromberg, B. J., Profeta, G. A., Andrews, J. J., and Walter, R. J. (1989). Use of a laser-induced force trap to study chromosome movement on the mitotic spindle. *Proc. Natl. Acad. Sci. U.S.A.* **86,** 4539–4543.
13. Ashkin, A., and Dziedzic, J. M. (1989). Internal call manipulation using infrared laser traps. *Proc. Natl. Acad. Sci. U.S.A.* **86,** 7914–7918.
14. Tadir, Y., Wright, W. H., Vafa, O., Ord, T., Asch, R. H., and Berns, M. W. (1989). Micromanipulation of sperm by a laser generated optical trap. *Fertil Steril.* **52,** 870–873.
15. Bonder, E. M., Colon, J., Dziedzic, J. M., and Ashkin, A. (1990). Force production by swimming sperm-analysis using optical tweezers. *J. Cell Biol.* **111,** 421A.
16. Ashkin, A., Schütze, K., Dziedzic, J. M., Euteneuer, U., and Schliwa, M. (1990). Force generation of organelle transport measured in vivo by an infrared laser trap. *Nature* (*Lond.*). **348,** 346–352.
17. Block, S. M., Goldstein, L. S. B., and Schnapp, B. J. (1990). Bead movement by single kinesin molecules studied with optical tweezers. *Nature* (*Lond.*) **348,** 348–352.
18. Sato, S., Ohyumi, M., Shibata, H., and Inaba, H. (1991). Optical trapping of small particles using 1.3 μm compact InGaAsP diode laser. *Optics Lett.* **16,** 282–284.
19. Gordon, J. P. (1973). Radiation forces and momenta in dielectric media. *Phys. Rev. A.* **8,** 14–21.
20. Ashkin, A. (1978). Trapping of atoms by resonance radiation pressure. *Phys. Rev Lett.* **40,** 729–732.
21. Gordon, J. P., and Ashkin, A. (1980). Motion of atoms in a radiation trap. *Phys. Rev. A.* **21,** 1606–1617.
22. Roosen, G., and Imbert, C. (1976). Optical levitation by means of 2 horizontal laser beams–theoretical and experimental study. *Physics. Lett.* **59A,** 6–8.
23. Ashkin, A., and Dziedzic, J. M. (1971). Optical levitation by radiation pressure. *Appl. Phys. Lett.* **19,** 283–285.
24. Ashkin, A., and Dziedzic, J. M. (1975). Optical levitation of liquid drops by radiation pressure. *Science* (*Wash. DC*). **187,** 1073–1075.
25. Ashkin, A., Dziedzic, J. M., Bjorkholm, J. E., and Chu, S. (1986). Observation of a single-beam gradient force optical trap for dielectric particles. *Optics Lett.* **11,** 288–290.
26. Ashkin, A., and Dziedzic, J. M. (1989). Optical trapping and manipulation of single living cells using infra-red laser beams. *Ber. Bunsen-Ges. Phys. Chem.* **98,** 254–260.
27. Wright, W. H., Sonek, G. J., Tadir, Y., and Berns, M. W. (1990). Laser trapping in cell biology. *IEEE* (*Inst. Electr. Electron. Eng.*) *J. Quant. Elect.* **26,** 2148–2157.
28. Born, M., and Wolf, E. (1975). Principles of Optics. 5th ed., pp. 109–132. Oxford: Pergamon Press.
29. Mansfield, S. M., and Kino, G. (1990). Solid immersion microscope. *Appl. Phys. Lett.* **57,** 2615–2616.

30. Richards, B., and Wolf, E. (1959). Electromagnetic diffraction in optical systems. II. Structure of the image field in an aplanatic system. *Proc. R. Soc. London. A.* **253,** 358–379.
31. Ashkin, A., and Dziedzic, J. M. (1973). Radiation pressure on a free liquid surface. *Phys. Rev. Lett.* **30,** 139–142.
32. Hori, M., Sato, S., Yamaguchi, S., and Inaba, H. (1991). Two-crossing laser beam trapping of dielectric particles using compact laser diodes. Conference on Lasers and Electro-Optics, 1991 (Optical Society of America, Washington, D.C.). *Technical Digest* **10,** 280–282.
33. Sato, S., Ishigure, M., and Inaba, H. (1991). Application of higher-order-mode Nd:YAG laser beam for manipulation and rotation of biological cells. Conference on Lasers and Electro-Optics, 1991 (Optical Society of America, Washington, D.C.). *Technical Digest* **10,** 280–281.
34. Ashkin, A., and Dziedzic, J. M. (1985). Observation of radiation pressure trapping of particles using alternating light beams. *Phys. Rev. Lett.* **54,** 1245–1248.
35. van de Hulst, H. C. (1981). Light Scattering by Small Particles, pp. 114–227. New York: Dover Press.
36. Kim, J. S., and Lee, S. S. (1983). Scattering of laser beams and the optical potential well for a homogeneous sphere. *J. Opt. Soc. Am.* **73,** 303–312.

CHAPTER 2

Basic Laser Tweezers

Ronald E. Sterba and Michael P. Sheetz
Department of Cell Biology
Duke University Medical Center
Durham, North Carolina 27710

I. General Description

For many cell biology laboratories, a basic laser tweezers system capable of picking up cells, beads, or other microscopic objects and placing them at appropriate places is sufficient. This basic system would include a low-power infrared (IR) laser, mirrors, and focusing lenses, which can be added to a good research microscope for \$10,000–\$20,000. The major limitations of the basic system are the laser power and stability needed for force measurements and the proper laser beam configuration required for trapping small particles.

For more sophisticated measurements of force, the basic system needs mechanical stability, sufficient laser power (maximally 0.5–1.0 W at the sample) in a beam of uniform intensity, and a precise analysis system tuned to the time scale of the system under study. The cost for a force measurement system increases

METHODS IN CELL BIOLOGY, VOL. 55
Copyright © 1998 by Academic Press. All rights of reproduction in any form reserved.
0091-679X/98 \$25.00

to $30,000–$40,000 because of the costs added by a more powerful laser, a piezoelectric stage, an antivibration table, and a video image processing system. We describe here the factors involved in adding a tweezer system to a research video microscope capable of force measurements.

II. Microscope

Any high-quality video microscope with an epi-illumination port can have laser tweezers added to it, but fluorescence microscopy is compromised (Figs. 1 and 2). For simultaneous fluorescence and tweezers microscopy, an inverted microscope with a bottom port is commonly used (see Figs. 3 and 4). There are three critical considerations for the microscope: (*a*) a dichroic mirror to reflect the IR laser beam and pass the video light, (*b*) a high numerical aperture (NA) objective (1.0 or higher NA), and (*c*) the stability of the stage and the coupling between the video camera and the microscope. We address each of these issues in order.

Because the light path below the objective lens is shared between the laser trapping beam and the imaging transillumination, a dichroic mirror must be mounted below the objective. This dichroic mirror must reflect the laser light into this path while letting the transillumination light pass through to the camera. Custom dichroic mirrors are available for the IR region from several optics firms

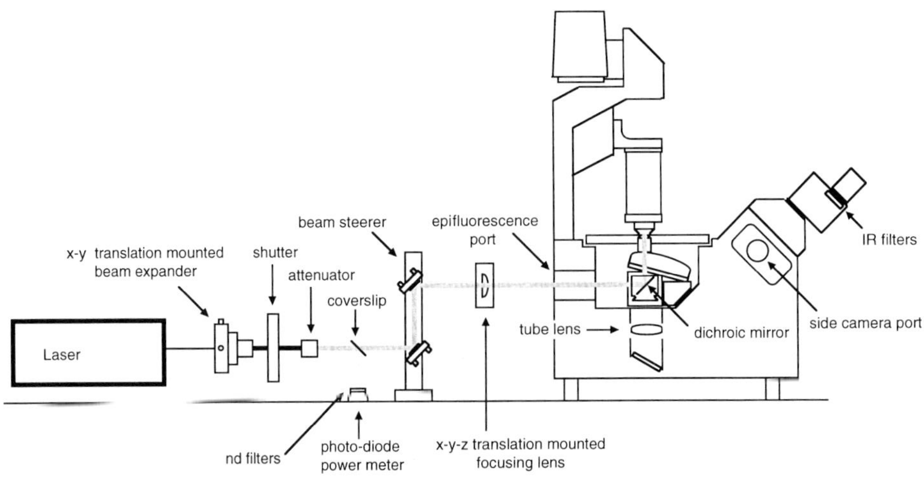

Fig. 1 Basic trapping system with shuttering, attenuation, and power measurement. The focusing lens is generally 75–200 mm focal length (plano-convex) and placed 50–75 cm from the back aperature of the objective.

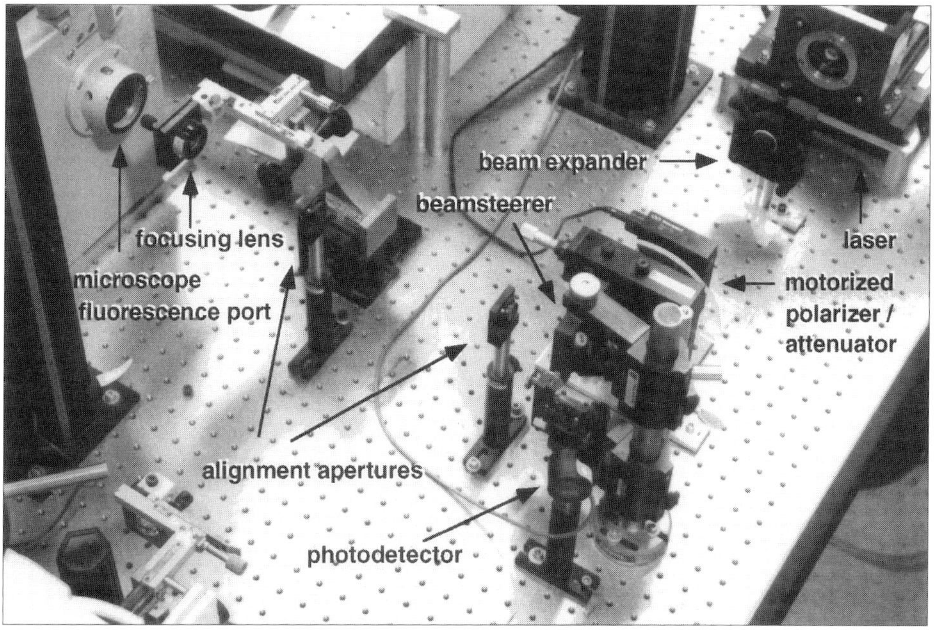

Fig. 2 Basic laser trap system using epifluorescence port.

(Chroma Technology Corp., Omega Optical Inc., Brattleborro, VT) and can be placed directly into a conventional fluorescence dichroic holder. Because the dichroics are somewhat variable, it is useful to obtain the reflectance and transmission characteristics for the actual dichroic that is purchased. The wavelength of maximum reflectance should match the wavelength of the laser light. The wavelength dependence of the transmission of light by the dichroic should be known to avoid loss of transmitted light. For simultaneous fluorescence and tweezers microscopy, a dichroic holder should be machined for the bottom port of the microscope. This dichroic holder should reflect all visible light to the camera and pass the laser light into the bottom port.

Using an objective lens with an NA of 1.0 or greater focuses the laser light at a sufficient gradient angle to form an effective trap (Svoboda and Block, 1994). Most objectives with a high NA can be used for tweezers construction, but we know of none currently that have been designed for the IR wavelengths. As a result of achromatic aberrations, most tweezers will lose trapping power with distance from the glass surface. Often particles escape the trap when the focal point is over 20 μm from the glass surface. With the increased importance of both 2-photon and laser tweezers applications, it is likely in the future that some

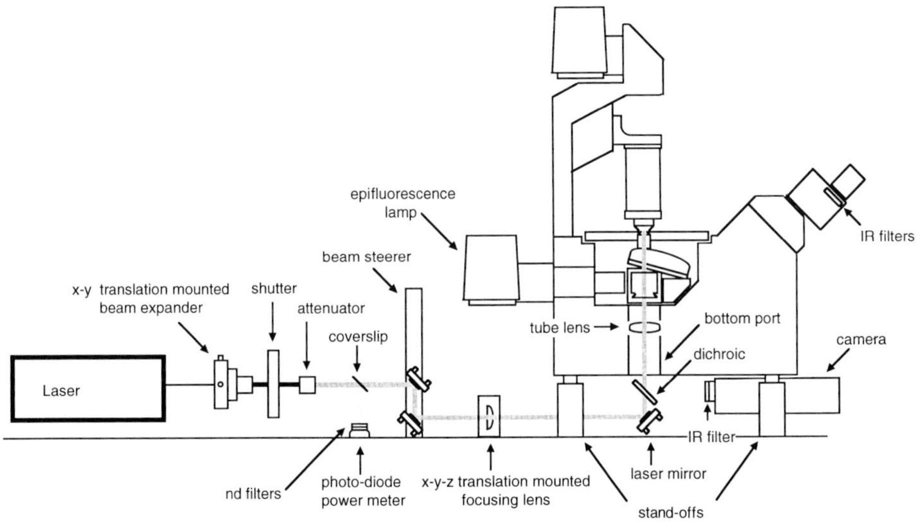

Fig. 3 Trapping system using bottom port for laser access allowing use of epifluorescence lamp.

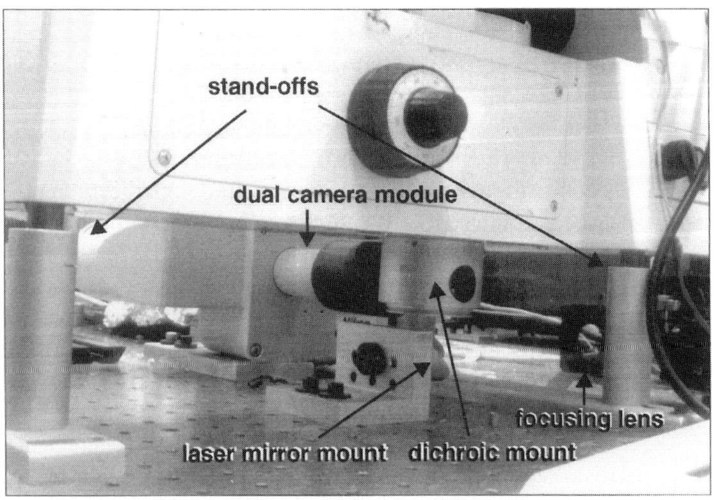

Fig. 4 Laser trap system using bottom port.

objectives will be designed for the IR range or at least will be tested for achromatic aberrations in the IR wavelengths.

There have been few cases in which objectives have been damaged by the high levels of IR light (1–3 W) used in some laser tweezers. This is because approximately half of the laser light is absorbed in the objective. To avoid possible damage and for safety reasons, always align the beam at low power levels. There are many cases of objective damage from vaporizing plastic apertures. Some objectives have apertures made of black plastic, which absorbs laser light and can vaporize if strongly illuminated. Once the black plastic is deposited on the glass surfaces of the objective, it degrades the image and is difficult to remove (black anodized metal apertures do not suffer the same problem).

For most tweezers applications, stability of the stage and the video camera-microscope coupling are critical. Any vibration of the stage, particularly vertical instability, will cause particles to be lost from the tweezers. Furthermore, thermal stage drift, mechanical backlash, and hysteresis from piezoelectric positioners will greatly increase investigator frustration. Motorized mechanical stages can move over distances of several centimeters and are useful for placing objects in the specimen plane, but these have large step sizes of 250 nm or larger. For very small precise movements involved in force measurements a piezoelectric stage can be used. For the force on any trapped particle to be measured, the position of the particle should be measured to within several nanometers (the linear portion of the force vs. displacement plots is 100–300 nm maximally). If there is any movement of the video camera relative to the microscope, then small displacements can not be reliably measured.

III. Laser

The selection of the laser system should take into consideration requirements of wavelength, laser power, beam pointing stability, beam mode quality, and noise. Also, older lamp pumped systems can require special three-phase high-voltage electrical supplies as well as a high-volume water supply for cooling. In general, in the IR spectrum biological material is more transparent at longer wavelengths, and water is more transparent at shorter wavelengths. A range of wavelengths between 780 and 1100 nm falls between these regions of relatively high absorptions (Svoboda and Block, 1994). Chapter 2 deals with issues of wavelength in greater detail.

To form a stable laser trap, the laser light should be a continuous-output low-noise beam that can be focused to as small a focal point as possible. The diameter of smallest focal point will be on the order of the wavelength of the light and is described as a diffraction-limited spot. To form a diffraction limited spot in the sample plane, the laser beam must be a single Gaussian peak with minimal side bands. This single peak emission is known as single mode or TEM00. A laser with a continuous output is described as a continuous-wave or CW laser. Laser systems which are not of the continuous-wave type are described as Q-switched or mode-locked lasers and have a high frequency pulsing output.

Laser systems in this range are currently available commercially in four different types. These lasers are summarized in Table I (Svoboda and Block, 1994). The first type is the solid-state CW Nd:YAG which emits at a wavelength of 1064 nm. Similar to the neodymium:yttrium aluminum garnet (Nd:YAG), the Nd:yttrium lithium fluoride (Nd:YLF) emits at wavelengths of 1047 and 1053 nm. These lasers are available in relatively high powers (1–3 W). Solid state refers to the fact that the photon source is a solid crystal rod of YAG, which is the host material for the active element, Nd. The Nd ions emit at 1064 nm when their electrons are stimulated to a higher energy level by a "pump" energy source, then drop back to their original orbit. Older systems used arc lamps to pump photon energy into the YAG rods. These lamp-pumped systems require large cooling systems and have a typically noisy output. Their laser light output can have high frequency ripple noise of 20% or more. Current versions are diode laser pumped, remain stable to less than 1% ripple, and require less cooling.

The second laser system type is a CW tunable Ti:sapphire laser system. This system has the advantage of being wavelength tunable over a range in the near infrared (NIR) laser from 650 to 1100 nm. The Ti:sapphire laser uses a green argon or frequency-doubled Nd:YAG laser to pump a tit:sapphire crystal rod. The fluorescent output of the rod is frequency filtered in a lasing cavity to permit tuning of the output wavelength. These laser systems, which contain many optical components, are hard to align and maintain. Also, pointing and temporal stability are dependent on the stability of the pumping laser.

The third laser system type is the lower-power but inexpensive CW single-mode diode laser system. A forward-biased p-n junction will emit photons that lase in a reflective cavity. The lasing photons are directed by a wave guide constructed of layers of material with different refractive indexes and are emitted

Table I
Continuous-Wave Near-Infrared Trapping Lasers

Laser type	TEM00 power	Wavelength	Price range
Solid state	100 mW–10 W		\$5000–\$40,000
ND:YAG		1064 nm	
ND:YLF		1047, 1057	
Ti:sapphire	2W	650–1100 nm Continuous Tunable	\$20,000–\$30,000, not including argon or ND:YAG
Semiconductor laser diode	5–250 mW	780–1020 nm	\$50–\$1000, not including power supply
Semiconductor MOPA laser (SDL Inc.)	1 W	985 nm	\$10,000, not including power supply

from one end of the wave guide. Because of manufacturing limitations of CW laser diodes, the emission divergence is elliptical but can be circularized and collimated with an aspheric lens.

The fourth laser system type useful for laser trapping is a special type of semiconductor laser system described as a monolithically integrated master oscillator/power amplifier (MOPA) manufactured by SDL Inc. of San Jose, California. It is a CW single-mode collimated laser system integrated with a thermoelectric cooling system and emitting at a single frequency of 985 nm at a power of 1 W. It has advantages of being relatively small, stable, and easy to maintain.

The laser light delivered to the sample plane will be a fraction of the laser source power. Microscope objectives absorb as much as half or more of the laser light entering the back aperture. Most high NA objectives contain several individual lenses that are antireflection coated for optimal visible and ultraviolet transmission. As a result, much of the IR light is internally reflected and absorbed in the objective. Microscope objective manufacturers typically suggest that laser powers of less than 1 W be used to reduce risk of damage due to heating of internal lenses and mounts. Optics used to steer and focus the laser light [e.g., mirrors, lenses, single-mode fibers, and (acousto optic modulators) AOMs] will also attenuate laser light.

IV. Optics and Layout

Mechanical stability is a critical concern in the maintenance of a trapping system. Therefore, the optical trapping system should be set up on a vibration isolated optical table. The laser and microscope should be set into position against mechanically fixed reference points on the table. Working with an inverted microscope is convenient because most of the access ports to the objective side of the optical path are relatively low and close to the table surface. The easiest and most straightforward optical path for the laser to enter an inverted microscope probably is through the epifluorescence port. A dichroic mirror reflective at the laser wavelength and transparent to the imaging illumination can be mounted in the 45° mirror position to reflect the laser into the objective.

When working with IR invisible laser light, it is important to wear laser safety goggles that block the specific laser wavelength. Because IR wavelengths greater than 800 nm are invisible to the human eye, it is important to use an infrared sensor card to visualize the beam as you set up the laser light path. These IR sensor cards will fluoresce at visible wavelengths when excited with IR laser light.

The laser light can be diverted into the microscope by use of mirrors or single-mode fiber optics. The easiest setup is a two-mirror beam steerer. A beam steerer is essentially two mirrors, each mounted on gimbals with fine angular adjustments. The gimbals, in turn, are mounted on a vibration-resistant post. The mirrors can be adjusted in tandem to change both the lateral placement of the beam in the objective back aperture as well as the beam angle into the objective. Laser-quality mirrors should be able to reflect 95% or more of light at the specific wavelength. The reflectivity of mirrors also will depend on the angle of incidence of laser light. In beam-steering applications, most reflections will be close to 45° incidence. In the case of plane-polarized light reflected at 45°, the reflectance spectrum of dielectric coatings can vary greatly if the electric field of the laser beam is parallel to the mirror surface (p plane) or perpendicular (s plane). Mirror manufacturers usually provide useful p-plane and s-plane reflectivity spectrum plots.

A laser beam expander should be used to magnify the laser's exit beam diameter to the objective lens back-aperture diameter. Beam expanders are commercially available in different magnifications with fixed entrance and exit aperture diameters. Many beam expanders have an adjustable focus control. Adjusting the focus will change the beam divergence and can be convenient for rough adjustments of the trap focus in the sample plane. For smaller more precise adjustments of the trap focus, an external lens can be used before the laser enters the microscope. It is important that both the beam expander and additional lenses be antireflection coated for the laser wavelength to ensure efficient laser transmission and to minimize stray laser light, which is an eye hazard and may also contaminate video camera images.

V. System Setup

There are many different methods for initially setting up and aligning the trap. The dichroic mirror surface designed to pass all the illumination light will still reflect a small percentage (1–5%) of this visible light. One technique is to use this small amount of reflected transillumination light to define the optical center of the microscope. By placing an oiled coverglass on the objective, then focusing and centering the partially closed condensor field aperture, an image of the aperture should be reflected out through the epifluorescence port. After mounting and centering the beam expander at the laser output, the beam-steerer mirrors can be adjusted to reflect this field aperture image back to the beam expander. Adjusting the beam-steerer mirrors to make the laser and illumination paths concentric will give a good approximate alignment. The beam alignment at the objective back aperture can be adjusted with the beam-steering mirrors and viewed by rotating the objective turret to an empty position. The laser light should be centered on and filling the back aperture. Clearing the space above the turret by tipping or removing the condensor will allow the laser light to project on the ceiling. The laser beam should be slightly divergent from the turret to the ceiling to put the trap focus close to the camera focal plane.

To view the trap in the sample plane, a coverslip mounted on a slide with 1-μm beads in water can be placed on the microscope. Without the IR camera filter in place, there should be enough light reflected from the coverslip water interface to image the reflected laser diffraction pattern. If the reflected laser light overdrives the camera, color filters passing the transilumination wavelength or neutral density filters can attenuate the laser light. Also, because the diffraction pattern of the laser usually has very high contrast, decreasing the camera gain and increasing black levels can flatten the image and make it possible to image more detail of the patterns. Adjusting the microscope focus above and below the inside surface of the coverslip should make the reflected diffraction pattern expand and contract radially to a μm-sized spot at the trap plane. If the radial pattern does not contract to a spot while the coverslip surface is in focus, the external laser-focusing lens must be added and adjusted to make the spot and the surface parfocal.

If the diffraction patterns are nonsymmetrical, the laser light is not hitting the glass–water surface orthogonally, and the angle of the laser as it enters the objective must be adjusted. A "walk-in" procedure using the two beam-steering mirrors while watching the laser pattern on a video monitor is one way to make the diffraction pattern and the trap symmetrical. If the trap is uneven in one axis, one beam-steering mirror should be adjusted to move the trap center off to one side of the video image. Then the trap center should be moved back with the other mirror. If the symmetry improves, this walk-in process should be repeated. If the trap symmetry decreases, the procedure should be reversed by moving the trap first in one direction with the opposite mirror.

Another rough alignment procedure uses a slide prepared with a high concentration of beads in water to image the symmetry of force of the trap. If the trap is misaligned, it will pull beads from one side and push them out in the opposite direction. The same walk-in procedure using the beam steerer while watching the bead manipulation can be used to make a symmetrical and stable trap.

VI. Alignment

For trapping cells the laser beam profile is often not critical, but it is essential for trapping small beads. A critical part of a trapping setup is the laser beam alignment through the objective lens and into the sample. To form a trap with even force in all directions, the cone of light that forms the trap must be very symmetrical. This is usually achieved by using a dual mirror beam steerer to make sure that as the laser enters the objective back aperture it is parallel with the center axis of the objective. Even a slight misalignment will result in an uneven trap that might grab particles from one side and push them out in the opposite direction.

To bring trapped objects into focus with the camera, the laser light must come to a diffraction-limited spot in the sample at the same focal plane as the camera. The focal height of the spot is effectively adjusted by changing the divergence of the beam as it enters the objective back aperture. Usually, the divergence can be adjusted with a beam expander or one or more additional lenses in the laser path. By controlling the laser focus height in small amounts around the camera focal plane, objects in the trap can be brought to different levels of focus with respect to the camera. Different bead sizes will be held at slightly different heights with respect to the camera focal plane. This means that the vertical trap position must be adjusted with different particle sizes.

VII. Translation

Optical trapping systems can be set up with one of two different methods for actuating movement in the sample plane. In a stationary optical trap the laser beam is steered into a fixed position in the sample plane. The laser focus point remains stationary while the sample is moved by either a motorized or piezoelectric stage. In a moving trap system the laser focus point can be moved in the sample plane by steering the trapping beam with movable mirrors, lenses, or acousto-optic modulators.

A stationary optical trap in a standard differential interference contrast (DIC) video image field will pull free objects and hold them in a stationary position with respect to the video field. By moving the microscope stage, the trapped objects remain stationary in the video field while cells or other biological material attached to the coverslip surface moves beneath the trapped particle. The trap can be positioned horizontally to capture objects in any part of the field and vertically to trap objects at or around the image plane of the camera. Stationary trapping systems work well for sorting or positioning microscopic particles and for force measurement work (Kuo and Sheetz, 1993).

A moving trap system allows sample movements as well as small movements of the trap position with respect to the video field. Moving trap systems allow very small and accurate movements of trapped particles by steering the laser beam with movable lenses, galvanometer mirrors, or acousto-optic elements. Moving trap systems are useful where high-speed movements of the trap are required as in feedback reenforcement systems (Finer, *et al.*, 1994). Moving traps are also useful in building a dual trap system so that one trap position can move independently of the other.

When trapping free particles or beads diffusing in solution, the bead must be positioned near the center of the trapping area. Because the laser light is filtered from the camera, there is no indication of the trap position unless a particle is held in the trap. The trap area in the video field can be marked on the video monitor screen directly with an ink marker. This will give an accurate indication of the lateral trap position and allow accurate positioning of free particles before trapping. In a moving trap system marking a "home" position allows initial trapping of particles.

VIII. Video Recording and Analysis

In general, a continuous video record of experiments by use of s-VHS or high-beta tapes is useful for documentation of timing and rare, revealing events. This recording can be cumbersome, and the proliferation of data tapes can be imposing. Nevertheless, the tapes are relatively cheap and can be analyzed with little loss of information (we typically get a resolution of 3–5 nm for bead tracking from tapes). Digital recording systems are an emerging competitive alternative that can store 30–120 min of video. Archiving such data is still expensive and time consuming, but simple data reduction schemes could improve that. The major value of the tapes is in screening a system, controlling for artifacts, and checking out useful parameters. Ultimately, a rapid method for extracting the few relevant parameters from the 250,000 bytes of data per image is needed. There is a rapid proliferation of video analysis systems that range in cost from about $4000 (NIH Image, which is useful for many applications, is free and uses a Macintosh system with a digitizing board) to tens of thousands of dollars for customized systems.

For measurements of force, the analysis is greatly simplified by using a stationary beam. Tracking the position of a trapped particle will give the force on that particle once the system has been calibrated. In the past, s-VHs tapes have been analyzed with personal computer (PC)-based routines (see Chapter 8). With the advent of digital video systems, anyone starting out is encouraged to invest in a digital recording system built around a PC that also analyzes the images.

IX. Accessories

It may be convenient to turn on the laser trap with a foot pedal or hand switch. Such a switch can operate a shutter to block the laser light from reaching the microscope. A black shutter capable of withstanding the laser beam's energy density or a reflective shutter that reflects the beam into a beam stop can be used. In the case of some open cavity lasers, the shutter can operate between the laser mirrors and block the lasing action.

Controlling the power of laser diodes, MOPA lasers and diode-pumped Nd : YAG or Nd : YLF lasers can be produced simply by controlling the operating current of the device. Limiting the operating currents of solid-state lamp-pumped lasers and argon-pumped Ti : sapphire lasers can affect the ripple noise and pointing angle of the laser output. Attenuation of these lasers can be accomplished with neutral-density linear wedge filter wheels or acousto-optic modulators. In the case of polarized lasers, the beam can be attenuated with rotating laser film polarizers or laser-grade Glan-Thompson prisms.

To indicate the power of variable laser power systems, the beam power can be detected and recorded on the video image. To observe and record the laser power, a piece of glass mounted at an angle in the laser path can reflect a small percentage of the full beam power to a photodiode detector. The detector can be protected from saturation by the full laser power with neutral-density filters. The photodiode's current must be converted to a linearly proportional voltage by an amplifier. This voltage can be digitized on a computer board and numerically displayed by using a computer-controlled video overlay (Horita, Inc.). Other commercially available dedicated devices digitize voltages and overlay this voltage on a video frame.

X. Summary

The basic information has been provided here for designing and building a laser tweezers system for force measurements. If force measurements are not required, then the considerations about the analysis system, a fine piezo stage, and stability are less important. For the initial alignment and characterization of the system, red blood cells provide an easily trapped sample. For a difficult test sample, the smaller latex beads (0.15–0.3 μm in diameter) are stable and easy to obtain. Anyone setting up laser tweezers is encouraged to see a working tweezers system and to compare samples with that system. Everyone has a different background, and there may be aspects critical for you that have not been discussed here. More sophisticated systems are described later in this book.

Appendix: Laser Tweezers Parts List

1. MOPA laser (SDL Inc., San Jose, CA)
2. Beam expander (Melles Griot, Irvine, CA; Newport Corp., Irvine, CA; Oriel Corp., Stratford, CT)
3. Beam attenuator (Melles Griot, Newport corp., Oriel Corp.)
4. Steering mirrors (Melles Griot, Newport Corp.: dielectric coating BD.2 or metallic coating ER.2, Oriel Corp., Thor Labs Inc.)
5. Focusing lens (generally 75–200 mm focal length (plano-convex) and placed 50–75 cm from the back aperture) (Melles Griot, Newport Corp., Oriel Corp.)
6. X-Y and X-Y-Z translation mounts (Melles Griot, Newport Corp., Oriel Corp., Thor Labs Inc., Newton, NJ)
7. Dichroic mirror (Chroma Technology Inc., Omega Optical Inc., both in Brattleborro, VT)
8. Infrared filters (Melles Griot, Newport Corp., Oriel Corp.)
9. Shutter (Melles Griot, Newport Corp., Vincent Assoc., Rochester, NY)
10. Photodiode power meter (Newport Corp., Thor Labs Inc.)
11. Infrared detector card (Quantex Corp., Rockville, MD)
12. Laser safety goggles (Uvex Corp., Fürth Germany)
13. Piezoelectric stage (Polytec Optronics Inc., Costa Mesa, CA; Wye Creek Instruments, Frederick, MD)

References

Finer, J. T., Simmons, R. M., and Spudich, J. A. (1994). Single myosin molecule mechanics: Piconewton forces and nanometre steps. *Nature* **368,** 113–118.

Kuo, S., and Sheetz, N. P. (1993). Force of single kinesin molecules measured with optical tweezers. *Science* **260,** 232–234.

Svoboda, K., and Block, S. M. (1994). Biological applications of optical forces. *Annu. Rev. Biophys. Biomol. Struct.* **23,** 247–285.

CHAPTER 3

A Simple Assay for Local Heating by Optical Tweezers

Scot C. Kuo

Department of Biomedical Engineering
The Johns Hopkins University
Baltimore, Maryland 21205

I. Introduction

Although reasonably transparent, water has measurable absorption at the near-infrared wavelengths typically used by optical tweezers (λ = 1064 nm). At this wavelength, the absorption coefficient of water is only $\varepsilon = 5.5 \times 10^{-4}$ cm^{-1} ($\varepsilon = [4\pi k(\lambda)]/[2.303\ \lambda]$, where $k(\lambda)$ is the absorptive index from Hale and Querry (1973)), hence absorbing less than 0.13% in a typical chamber 100 μm deep. However, cellular applications of optical tweezers often require >100 mW of laser power, and the consequent >0.13 mW absorbed will increase the aqueous temperature ~1°C/100 mW. In cylindrical cells, such as dissociated outer hair cells from the mammalian ear, this increased temperature causes a rapid, reversible 0.5–2% elongation (100–400 mW) of cell length that is *independent* of the optical forces (LeCates *et al.*, 1995). Such elongations are comparable to those induced by electomotility of these outer hair cells. Irradiation within a 50-μm field-of-view radius of an outer hair cell caused equivalent elongation, thus excluding any optical effects from the ~1-μm laser focus. Because most cells will not exhibit such an obvious response to heating, more subtle temperature effects require an independent method to estimate local heating by optical tweezers.

METHODS IN CELL BIOLOGY, VOL. 55
Copyright © 1998 by Academic Press. All rights of reproduction in any form reserved.
0091-679X/98 $25.00

A number of methods exist for monitoring the local temperature near the field of observation. Microthermocouples (e.g., Omega Corp) can be inserted near the field of view, but accidental irradiation of the thermocouple by the optical tweezers laser can cause temperature jumps more than 500°C, which vaporize the local media. Interestingly, transferrin particles have similar "opticution" effects. Tromberg and colleagues have developed a very sensitive microfluorometric method to monitor the red shift as fluorescent lipid vesicles "melt" through their phase transition temperature (Liu *et al.,* 1994, 1995). Their measurements indicate a 1.1 to 1.5°C average rise in temperature for 100 mW irradiation. Although less sensitive, a low-melting wax was used by Berg and Turner (1993) to characterize electrical heating of their microscopic specimens. We adapted the wax melting procedure for optical tweezers.

II. Methods

The butyl ester of stearic acid (99%) was purchased from Sigma (S-5001). Its melting point should be 27.5°C (CRC Press, 1987), but we measured its thermal properties using a Seiko differential scanning calorimeter (Model DS200). Empirically, we determined that it required 109 mJ/mg at 26.2°C to melt the butyl stearate. The reduction in melting point probably reflects impurities in the commercial specimen.

Test specimens were prepared by heating the butyl stearate to 37°C and placing 2 μl droplets on coverslips, where droplets quickly solidified. Smaller droplets were generated by incubating coverslips in a 37°C incubator and then "streaking" the droplet with a pipet tip. Microdroplets (~50 μm diameter, 1–3 μm thick) formed along the "streaks," which were used as test targets. After cooling, coverslips are assembled with the cell medium (phosphate-buffered saline for outer hair cells) on observation chambers. Typically, we used Scotch double-stick tape (75 μm thick) to form the chamber on a standard glass microscope slide (1″ × 3″). Wax melting was monitored using video-enhanced DIC microscopy (Kuo *et al.,* 1991), and the optical tweezers used a Santa Fe Nd:YAG laser (Model C-140, 10W, 1064 nm) through a Zeiss 100 × 1.3 Plan Neoflaur oil-immersion objective.

III. Results

With video-enhanced DIC microscopy, butyl stearate melting is very obvious because the surface tension of the wax causes it to "retract" from the glass coverslip. Starting at room temperature (22°C), the minimum laser power to melt the wax was ~250 mW at the specimen, with a small 1.5-μm spot melting within 0.5 sec. The minimal power to melt the wax did not vary significantly within 50 μm of the wax, which is consistent with the high thermal conductivity of water. Because the initial melt is ~5 pg, at least 0.5 nJ is required for melting, and the aqueous temperature must reach 26.2°C within 0.5 sec. In principle, heating the specimen with an air enclosure should reduce the laser power needed to melt the wax. The wax-melting method estimates 1.7°C/100 mW at the specimen. Although it is difficult to *directly* measure the optical transmission of high-NA objectives, the strength of this approach is that the laser power *entering* the high-NA objective is readily quantifiable, providing an empirical characterization of a particular optical tweezers set up.

Because most optical tweezers have been custom-built, if not customized for particular applications, estimates of aqueous heating are likely to vary between apparatuses. We present a simple, direct method to estimate the local heating by optical tweezers that we hope will be useful to all practitioners of the technology.

References

Berg, H. C., and Turner, L. (1993). Torque generated by the flagellar motor of *Escherichia coli*. *Biophys. J.* **65,** 2201–2216.

CRC Press (1987). *CRC Handbook of Chemistry & Physics*, 68th Ed., Boca Raton, Florida: Chemical Rubber Company Press, p. C-495.

Hale, G. M., and Querry, M. R. (1973). Optical constants of water in the 200-nm to 200-μm wavelength region. *Appl. Optics* **12,** 555–563.

Kuo, S. C., Gelles, J., Steuer, E., and Sheetz, M. P. (1991). A model for kinesin movement from nanometer-level measurements of kinesin and cytoplasmic dynein and force measurements. *J. Cell Sci. Suppl.* **14,** 135–138.

LeCates, W. W., Kuo, S. C., and Brownell, W. E. (1995). Temperature-dependent length changes of the outer hair cell. Association for Research in Otolaryngology, 1995 Midwinter Meeting. Abstract 622.

Lui, Y., Cheng, D. K., Soneck, G. J., Berns, M. W., Chapman, C. F., and Tromberg, B. J. (1995). Evidence for localized cell heating induced by infrared optical tweezers. *Biophys. J.* **68,** 2137–2144.

Liu, Y., Cheng, D. K., Soneck, G. J., Berns, M. W., and Tromberg, B. J. (1994). Microfluorometric technique for the determination of localized heating in organic particles. *Appl. Phys. Lett.* **65,** 919–921.

CHAPTER 4

Reflections of a Lucid Dreamer: Optical Trap Design Considerations

Amit D. Mehta, Jeffrey T. Finer, and James A. Spudich

Department of Biochemistry
Stanford University
School of Medicine
Stanford, California 94305

I. Introduction

The optical trap technique can be used to constrain and move small particles in solution using a light microscope and laser beam. Trapping size scales and sensitivity are well suited for studying the mechanical properties of single cells, organelles, and even molecules. Here, we describe considerations involved in the planning and implementation of an optical trapping microscope for high-resolution force and displacement measurements of trapped particles. Although the concerns are general, we describe them in the context of our experiments, which involve optical trapping of beads attached to single actin filaments. The filaments are then moved close to surfaces sparsely decorated with myosin mole-

METHODS IN CELL BIOLOGY, VOL. 55
Copyright © 1998 by Academic Press. All rights of reproduction in any form reserved.
0091-679X/98 $25.00

cules. These molecules will bind to and move the actin filament, allowing measurement of their mechanical properties at the single molecule level. We observe these beads with nanometer resolution, use active feedback loops to suppress bead diffusion by rapid trap deflection, and observe the specimen by using brightfield and fluorescent imaging simultaneously.

The issues discussed include choice of a trapping laser, design of the optical layout, imaging of the trapping plane, high-resolution position detection of the trapped particle, use of negative feedback to further constrain movement of trapped particles, calibration of the trap parameters, and analysis of force and displacement measurements.

II. Choice of Trapping Laser

Single-molecule measurements of biological motor dynamics require detection of nanometer displacements and piconewton forces with millisecond resolution (Finer *et al.*, 1994; Svoboda *et al.*, 1993). Such experiments have used trapping beams from neodymium (Nd)-doped : yttrium lithium fluoride (Nd : YLF) or Nd-doped : yttrium aluminum garnet (Nd : YAG) lasing crystals pumped by a diode laser. Such lasers provide the requisite stability, wavelength, and beam quality.

One consideration in choosing a laser involves the wavelength of light. The optimum wavelength depends on the size of the object to be trapped. Trap stiffness, or restoring force generated per particle displacement from trap center, is usually the parameter of interest. In general, the smallest possible beam waist will approximate the wavelength of laser light, and the strongest trap stiffness occurs for particles of the same size as the waist. Trapping stiffness drops sharply as particle size falls below this level but falls modestly as the size is increased (Simmons *et al.*, 1996).

A second concern is avoidance of optical damage to biological samples. In pioneering work, Ashkin and colleagues found that the argon green line at wavelength 514.5 nm caused trapped bacteria to burst at modest power levels (Ashkin *et al.*, 1987). In subsequent experiments, these researchers made use of an infrared Nd : YAG laser at wavelength 1064 nm, sufficiently far from protein and water absorption peaks to allow study of biological samples. Thus far, Nd crystal lasers remain the most widespread and best characterized (Ashkin *et al.*, 1987; Ghislain *et al.*, 1994; Simmons *et al.*, 1996). However, a tunable Ti : sapphire laser operating at 700-nm wavelength, further from water absorption lines than 1064 nm, provides a stronger trap for a given power and reduces laser-induced cell damage relative to Nd : YAG (Berns *et al.*, 1992). Moreover, sensitive measurements involving proteins or DNA attached to optically trapped beads have been performed using diode lasers in the 800 nm range (Smith *et al.*, 1996; Wolenski *et al.*, 1995).

In general, diode lasers are inexpensive, compact, and available at high-output power levels. However, they are easily destroyed by electrical transients and must be protected using specialized power supplies or bypass circuitry to shunt these transients as well as electrostatic shielding, especially if arc lamp discharges occur in close proximity. Moreover, the output beam from the diode element tends to be highly divergent and astigmatic, requiring care in collecting and collimating the output light for use in trapping. This problem is likely to fade in the near future with the advent of collimating microlenses that can be integrated with the basic diode package to provide a circular, diffraction-limited beam.

The quality of the output beam can be relevant, depending on experimental requirements for trap stability and linearity. Measurement of nanometer displacements and piconewton forces require nanometer position stability of the trap in the specimen plane. Necessary beam pointing stability is provided by the better diode-pumped solid-state lasers compared with older flashlamp-pumped versions. This can be a significant concern for precision measurements because microradian shifts in the beam at the laser output coupler cause nanometer movements of the trap in the specimen plane. The best solid-state lasers are stable within a microradian for about half a minute, with relatively negligible beam direction noise at higher frequencies. Moreover, a continuous flow of water is required to cool but flashlamp-pumped lasers and diodes that are used to pump high power (over 1 W) solid-state lasers. This introduces vibration into the trapping microscope, normally built upon a vibration isolation table. Thus far, this has not prevented experiments at the previously described resolution levels. Additionally, the latest solid-state lasers couple the diode-pumping beam into the main lasing cavity via an optical fiber, thus allowing the water-cooled diode laser to be placed in a remote location.

The optical trap seems fairly tolerant of defects in transverse mode quality. A fairly linear trap can be generated on a nanometer scale as long as most of the light is in a symmetric Gaussian mode. Single-molecular motor measurements have used lasers with no more than 80 to 85% of the light in the TEM^{00} symmetric Gaussian mode (the transverse electromagnetic wave supported by the cavity of the zeroth order in both transverse dimensions). More recent solid-state lasers have used novel pumping geometries to restrict higher order TEM modes to less than 5% of laser output power, but it remains unclear whether these marginal improvements will have a notable effect on trap quality. Additionally, higher order transverse modes can be removed from a "dirty" beam by passage of the light through a single-mode optical fiber or by focusing the beam through a wavelength-size pinhole. Thus far, trap linearity on nanometer scales has not been required in most applications.

The laser power level requirement depends on the size and dielectric properties of the trapped objects. For instance, single-molecular myosin step measurements have involved trapping of 1-μm-diameter polystyrene beads, in which the bead size is chosen to match the trapping beam wavelength: 12 mW from a 1047 nm Nd:YLF measured just before a 1.4-NA, 63× objective yielded traps with a stiffness of 0.02 pN/nm. Trap stiffness greater than this scales linearly with laser power. In such experiments, the weak traps are used to ensure that they are more compliant than a single motor protein molecule. In constructing a strong trap (0.15 pN/nm) for use in precise three-dimensional position measurements, Ghislain, *et al.* (1994) used 60 mW from a 1064 nm Nd:YAG measured after focus through a 63× objective with a numerical aperture (NA) of 1.25. Perkins *et al.* (1994) used 100 mW from a 1064 Nd:YAG measured at the focal point to pull an attached DNA molecule through a solution of entangled polymers to demonstrate reptation of the DNA. Berns *et al.* (1992) took 200–500 mW measured before the objective from a tunable Ti:sapphire laser set to 700-nm wavelength to pull and rotate chromosomes at different places in the mitotic spindle. However, a Nd:YAG laser at 1064 nm used in the same experiment could be operated only below 340 mW to avoid apparent laser damage to organelles. Ashkin *et al.* (1990) used 220-mW lasers to arrest moving mitochondria and 30- to 110-mW lasers to slow them down. The power measured before the objective often can be far less than the power required from the laser, especially if many optics and/or electro-optic spatial light modulators are included in the beam path. In an application requiring very high tension levels, Smith *et al.* (1996) used two counterpropagating beams at 800-nm wavelength to create 70 pN of force on a 1-μm-diameter polystyrene bead.

III. Optical Layout

Multiple traps require many independently steered trapping beams. In some applications, many traps may be necessary to constrain large, irregular objects. In other applications, two separated beams are used to trap different particles independently. If the optical power is sufficient, a single beam can be split into components by a polarizing beam splitter. A half-wave plate can be positioned before the beam splitter to rotate the polarization of light and thus change the fraction of light in each of the split beams. Alternatively, a beam deflector using an acoustic–optic element or a piezoelectrically driven mirror can be used to deflect a trapping beam rapidly between multiple positions, thus effectively creating many separate traps (Visscher *et al.*, 1993). Once the two trapping beams are separated, they must be steered independently and expanded. For these purposes, the beam must be fairly well collimated. A lens inserted before the splitting corrects the slight divergence of light emerging from the lasing cavity.

An optical trap requires filling the back aperture of a high-numerical-aperture microscope objective with a parallel laser beam. Adjusting this beam's angle of incidence on the objective will laterally shift the laser focal point within the specimen plane. This is the goal if one seeks to move the trapping point in the specimen plane without significantly changing the trap strength.

The following description of the optics in our double-beam laser trap is essentially backward, beginning with the focused beam and moving from there to the laser source. As illustrated in Fig. 1, in which only one of the two beams is shown

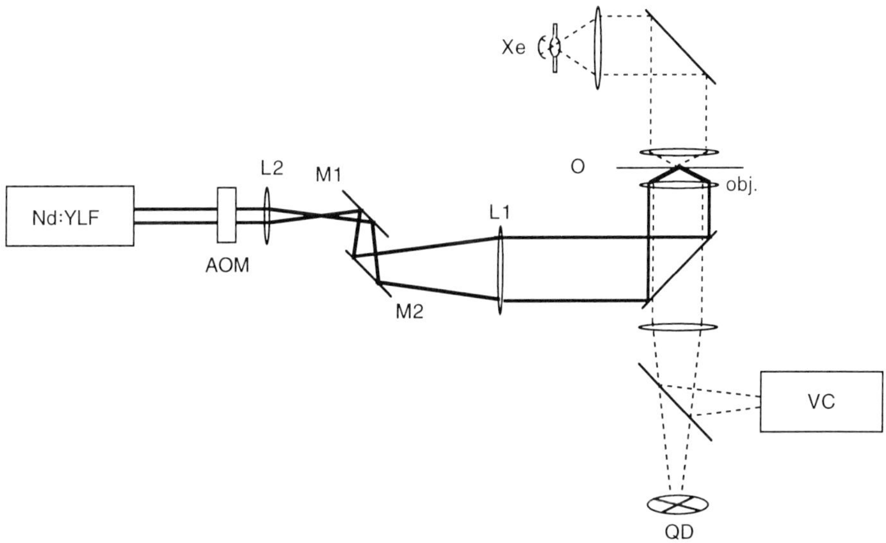

Fig. 1 Schematic of optical trapping and imaging system. Solid lines reflect the Nd:YLF laser beam, whereas dotted lines reflect light from the xenon arc lamp. Optics include lenses L2 and L1 for beam expansion; mirrors M1 and M2 for slow, manual steering; acousto-optic element AOM for rapid, electronically controlled trap deflection; and a microscope objective for bringing the beam to a diffraction-limited focus. Illuminating light from the arc lamp is split between a CCD camera and a quadrant photodetector to provide bead position information with millisecond and nanometer resolution.

for simplicity, the beam is focused to a diffraction-limited waist by the objective. As mentioned previously, just before the objective, the beam must be expanded, collimated, and have an adjustable angle of incidence to enable lateral movement of the trap. The beam is collimated if laser light diverges from a point in the rear focal plane of the lens L1. As long as the beam underfills L1, a larger focal distance will increase the size of the collimated beam, resulting ultimately in a steeper optical gradient and stronger trap. Finally, the lateral position of the beam before L1, perpendicular to the propagation axis, must be under user control. To this end, mirrors M1 and M2, which are used to steer the beam to L1, are placed on motorized translation stages and driven by joystick. By simple geometric optics, a lateral shift of the beam incident upon L1 is optically equivalent to an angular shift of the beam after L1 and thus a lateral shift of the trap in specimen plane O. The shift in mirror M1 or M2 produces in the O plane a corresponding trap movement, smaller by a factor of f1/fobj. Because f1 is 750 nm, and fobj is 3 mm, this is a demagnification of approximately 250×. As an alternative to the use of mirrors, lens L2 can be shifted laterally to achieve similar beam steering.

Backtracking along the beam path, lens L2 is used to focus the beam to a Gaussian waist at the rear focal plane of L1. The beam will then diverge from this plane and, as mentioned earlier, be collimated by L1. Note that this is an approximate relationship, and the distance between L1 and L2 will change as the mirror positions are shifted to steer the trap. The trapping focal plane at the specimen is relatively insensitive to these minor perturbations. The transverse magnification of the beam caused by L1 and L2 is simply f1/f2. In our apparatus, 5-cm f2 and 75-cm f1 result in a magnification of 15×. An alternative method of transverse magnification involves a Gallilean beam expander, a diverging lens followed by a collimating, converging one (Block, 1990).

The motorized mirrors are used for slow trap displacements over many micrometers. The beam continues roughly to fill the back aperture of the objective as the mirrors are moved over several millimeters, thus preserving trap strength for displacements of approximately 10–15 μm in the O plane. In many applications, the trap position must be placed under fast and accurate electronic control. Electro-optic spatial light modulators can be used to create small, precise trap deflections on electronic time scales.

Our instrument employs acousto-optic cells (AOM) for this purpose. These cells consist of a transparent crystal into which acoustic distortion fronts can be launched by a piezoelectric transducer. AOM drivers typically include a voltage-controlled radio frequency oscillator that accepts either analog or digital input, and a high-voltage amplifier to drive this transducer in the radio frequency range. The acoustic waves propagate by inducing local distortions of the crystal lattice. These distortions cause a small change in the local refractive index, a phenomenon known as the acoustic-optic effect. The cell thus behaves as a thick, one-dimensional phase grating with periodicity set by the spatial frequency of the lattice distortions. The effective driving voltage spectrum is typically centered about some RF frequency, with a limited bandwidth about that frequency through which the cell can be driven. In a phenomenon known as the Bragg effect, the large size of the crystal relative to the distortion wave period introduces a weighting for the zero and one first-order diffraction peak at the expense of higher orders. In fact, strong deflection into a first diffraction order occurs when the beam angle of incidence has a particular value known as the Bragg angle. The crystal must be carefully aligned to optimize laser power transmission into the first diffraction order of choice. AOMs are available with very high first-order peak transmission efficiencies (95%) and high incident power damage thresholds (~10 W) from manufacturers including Isomet Corporation and Newport Electro-Optics System (NEOS). High transmission efficiency typically comes at a cost in variation of the transmitted power as a function of deflection angle. To steer the beam, one changes the signal input to the voltage controlled oscillator, resulting in a shift of the drive frequency that, in turn, shifts the period of the phase grating and the deflection angle of the first-order peak.

Two AOM cells must be combined to steer the trap in both dimensions of the specimen plane. The first crystal will pass an undeflected zero-order peak and a first-order peak in one particular direction, here called x. The second crystal will pass both of these peaks, as well as first-order deflections of each in the other direction, here called y. Thus, one creates an undeflected peak, a first-order peak in x only, a first-order peak in y only, and a first-order peak in both x and y. A beam stop then blocks the former three and allows the latter peak to pass. Thus, the beam ultimately used in the trap can be shifted in either direction perpendicular to its axis. The two crystals are placed just to either side of the rear focal plane of L1. The angular deflection caused by the AOMs is optically equivalent to a lateral shift of the beam between L1 and L2 and thus equals a lateral movement of the trap in the specimen plane.

Alternatively, one can use a two-dimensional piezoelectrically driven mirror to steer the beam quickly. This technique is used in most laser-scanning confocal microscopes. Relative to electro-optic devices, it is simpler to configure, but the response will have less bandwidth.

At this point, the two beams have been independently expanded and steered. The beams are brought into close proximity before focusing through the microscope objective. Our system uses a 63× oil immersion NA-1.4 objective (Zeiss, Oberkochen, Germany). This high numerical aperture results in sharp focusing of the trapping beam, which is essential because the gradient of optical intensity determines the trapping force. However, the working distance is quite small (200–300 μm), and the trap cannot be moved far beyond the 175-μm width of a typical microscope coverslip. Moreover, it becomes more difficult to trap particles at a depth greater than 20 μm in the solution cell because spherical aberration causes a blurring of the laser focus. The trapped particle will escape in the axial direction, first because the trap is weaker in that direction (Ghislain *et al.*, 1994), and second because the particle is already displaced from trap center along the beam axis as radiation pressure from the light counters the restoring force from the light intensity gradient. A low-NA objective can be used to increase the working distance, but at the cost of an even weaker, possibly ineffective trap. One can compensate by using two counterpropagating traps, both focused at the same point. The radiation pressure effects will cancel, making the trap much more effective in what is ordinarily its weakest direction (Smith *et al.*, 1996). Additionally, a water immersion objective of high NA (1.2 or 1.0) can be used for trapping deep in the flow cell because spherical abberations are much less a problem (R. M. Simmons, personal communication).

IV. Imaging

Once the laser is focused and the trap created, the trapping plane must be imaged. Our measurements require simultaneous observation of the specimen plane in bright field and fluorescence as well as nanometer resolution detection of trapped beads. In our instrument, the trapping beam-focusing objective is also part of an inverted microscope. Dark-field, phase, or Nomarski optics may be incorporated to increase image contrast and sharpness. However, for making precise, quantitative position measurements of intrinsically high-contrast objects, such techniques are unnecessary and often counterproductive. A trapping beam can be steered into a standard upright microscope, but such instruments are more prone to vibration, especially if massive attachments to support position detectors or cameras are mounted near the top. This problem is addressed by using an inverted microscope, assembled from only the essential components. The illumination source, a 75-W xenon arc lamp, is mounted on top of the microscope column, whereas the imaging optics, detectors, and cameras are mounted on the vibrationally isolated table directly. A 1.4-NA oil immersion condenser is positioned just above the specimen slide, and the inverted objective is placed below it.

Bright-field microscopes usually employ Kohler illumination, in which the light source is imaged in the back focal plane of the objective and each section of the specimen is illuminated by parallel rays coming from an extended region within the source. Although this reduces the effect of a spatially inhomogeneous light source on the image of the specimen, it does not optimize brightness of the light falling on the specimen. High-resolution position detection of a trapped particle requires intense light for reasons discussed later. Brightness may be optimized by critical illumination, in which the light source is simply imaged on the specimen. Typically, we simply adjust arc lamp collection optics and the condenser position to optimize image brightness. Because nanometer deflections of a micron-size particle are of interest, inhomogeneities of larger spatial dimensions can be tolerated. Weak light intensity, however, can be particularly problematic (see later). The optimal illumination scheme can vary depending on the application and the light source quality.

Many applications require visualization of fluorescent probes, sometimes in addition to the bright-field image. Simultaneous fluorescence and bright field requires care in spectrally isolating the two images from each other. Our experiments require detection of actin filaments, which are labeled using rhodamine–phalloidin. To prevent bright-field light from overwhelming the red fluorescent rhodamine emissions, a filter is used to block passage of light less than approximately 780 nm in wavelength. Note that the xenon arc lamp is the brightest incoherent illumination source given the 780- to 1000-nm spectral constraint (the 1000-nm high end constraint is due to deflection of the trapping beam, explained later and illustrated in Fig. 2). If there were no 780-nm low end constraint, a mercury arc lamp would deliver higher intensity levels for use in the bright-field image, although the xenon arc spectrum is better matched to available silicon photodiodes. Although these wavelengths are specific to a dye excited in green and emitting in red, the basic scheme and considerations generalize. The numbers need merely to be changed.

Fluorescent excitation is provided by a mercury arc lamp coupled to the microscope via an optical fiber in our experimental setup. An excitation filter centered at 540 nm with a bandwidth of approximately 35 nm is used to filter the light to provide a spectrally narrow green beam.

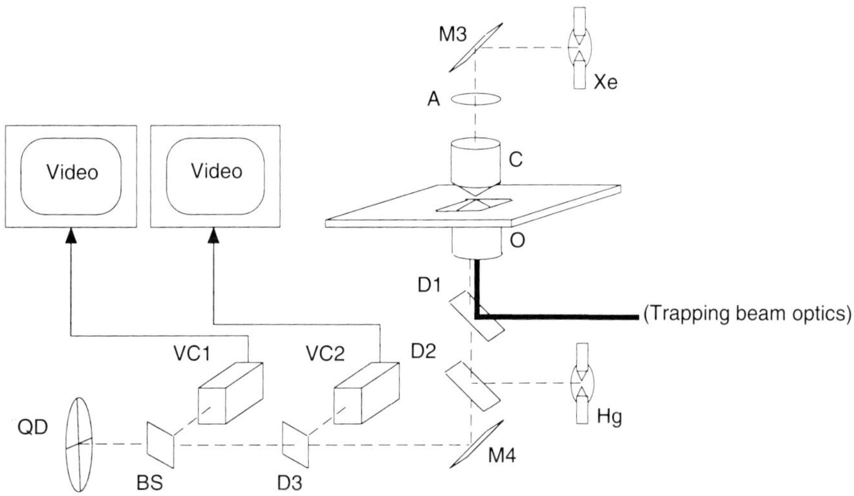

Fig. 2 Schematic of dichroic beam splitters positioned to allow both trapping and simultaneous bright-field and fluorescent imaging. Dichroic beamsplitter D1 reflects light with wavelength more than 1000 nm; D2 reflects light in the range of 520 to 560 nm; and D3 reflects light less than approximately 630 nm. All three dichroic beamsplitters operate at 45° incidence. Light from the xenon arc lamp (Xe) optically filtered to remove wavelength components less than approximately 700 nm, is deflected downward by mirror M3 and focused by the microscope condenser onto the specimen. Aperture A is used as a reference; the light is focused to optimize image brightness in a plane conjugate to A, allowing the experimentalist to find this plane again by moving the condenser to bring A into focus. This bright-field image light passes through all three dichroic beam splitters, with a small amount deflected by beam splitter BS and CCD camera VC1, and the remainder falling upon the quadrant detector QD. The trapping laser beam at 1047 nm will be deflected by D1 up into the microscope objective, which focuses it to a 1-μm waist in the specimen plane. Green fluorescence excitation light from the mercury arc lamp is deflected upward by D2 and focused through the objective onto the specimen. Downward red fluorescent emissions will pass through D1 and D2, then be deflected into SIT camera VC2 by the dichroic beamsplitter D3. Not shown are the fluorescent excitation filter, placed in front of the Hg lamp at normal incidence and allowing passage of a 30- to 40-nm-wide band centered around 540 nm, and the fluorescent emission filter, placed in front of VC2 at normal incidence and allowing passage of a 60- to 80-nm-wide band centered approximately 615 nm. Hence, the trapping beams and fluorescent excitation light are deflected upward into the objective, and the bright-field image light and fluorescent emissions are sent downward and deflected by mirror M4 into the imaging pathway, where the two images are separated by D3. Light greater than 1000 nm must also be blocked between D3 and BS because backscattered light from the trap will otherwise appear in the bright-field image.

Below the objective, dichroic beam splitters must be positioned to steer the trapping beam, image light, and fluorescent illumination in the proper directions. This scheme is illustrated in Fig. 2. Immediately below the objective, a dichroic filter (DI) deflects all light greater than 1000 nm and passes light less than this at a 45° incidence. The 1047-nm trapping beam, incident from the side, is deflected upward and into the objective. Illumination light of wavelength less that 1000 nm, coming down from above, will pass through this filter. Below this, a second dichroic beam splitter (D2) passes light of wavelength greater than approximately 570 nm and reflects light in the range of 520 to 560 nm. Green light coming from the mercury arc lamp will be deflected upward and focused through the objective into the solution cell, whereas image light, including both the red fluorescent emissions and the bright-field image, will pass through both dichroic beam splitters coming from above. A mirror is positioned at the bottom to deflect the light into an imaging pathway.

The fluorescent image is deflected into a silicon-intensified target (SIT) camera (VC2) using a dichroic beam splitter (D3), which reflects light of wavelength less than approximately 630 nm at a 45° incidence. Ordinary tube cameras have far too much dark noise to be useful for such low-light imaging, but intensified cameras, such as SIT, ISIT, or intensified CCD (ICCD), tend to provide requisite sensitivity for good image signal-to-noise ratio and produce images at video rate. Cooled, slow-scan CCDs do not provide video rate output but have very high sensitivity by use of on-chip integration to reduce electronic noise. Additionally, real-time digital image processing techniques can be used to enhance contrast or subtract background light levels to provide a clearer image.

Because the fluorescent signal is generally weak, the number of optical elements separating the specimen from the SIT camera must be minimized to optimize the light throughput. In addition to the microscope objective, one lens is positioned before the beam splitter to place an image of the trapping plane on the front end of the SIT camera. A 20× compensating eyepiece is placed before the camera to add magnification until a field approximately 100 μm in diameter can be seen in the image displayed.

Of the remaining light not deflected into the SIT, 10% or so is deflected into a video-rate CCD camera (VC1) using glass (BS) coated for antireflection on the rear side. This light provides a bright-field image of the field, set at approximately the same magnification as the fluorescent image. The aforementioned camera concerns are not relevant here because there is a very high level of light from the xenon lamp. Several lenses, both diverging and converging, must be placed in the path to ensure that an image of the specimen is in focus at the CCD detector array.

V. High-Resolution Position Measurement

The light not deflected into the CCD camera proceeds to sensitive light detectors. Thus far, the camera-generated images are useful for steering the trapped particles into an appropriate geometry, but do not have the spatial or temporal resolution to track single molecular motor movements. This resolution is achieved by using a lens and a 20× eyepiece to focus an enlarged image of the trapped particle on the quadrant photodetector. The basic idea is illustrated in Fig. 3. Magnification at this point is roughly 600× to 700×, so the image of a 1-μm bead fills most of the 1-mm-sided detector.

This detector is a specialized 4-element silicon photodiode array, with each element in one quadrant of the detector plane. Such detectors are not available in a wide variety of types of spectral sensitivities. Our system uses a detector with a peak sensitivity wavelength range of 800–1000 nm (S-1557, Hamamatsu). An image of the trapped bead, appearing as dark against a bright background is placed on the detector. If the image moves, say, upward, more light falls on the lower two elements and less upon the upper two. Thus, the lower two elements generate more current and the upper two less.

It is important to note that by such a method the diffraction-limited resolution can be exceeded, by a factor of 1000 to 10,000 or more. The diffraction resolution limit merely suggests that different specimen points closer than the limit will not be distinguishable because each appears in the image as convolved with the modulation transfer function (MTF) of the imaging system. In this case, where incoherent light from an arc lamp is used to illuminate the specimen, the MTF is the normalized autocorrelation function of the Fraunhofer diffraction pattern of the imaging optics exit pupil. This convolution kernel tends to be about as broad as the light wavelength, and will thus wash out details finer than the wavelength. However, in the situation described here, an image of the particle, or a dark silhouette of the particle convolved with the MTF, appears on the detector. If this silhouette moves by a tiny fraction of its total size, a change can be detected in the differential current output from the photodiode elements. Such considerations would apply even if the actual particle were considerably smaller than the wavelength of imaging light.

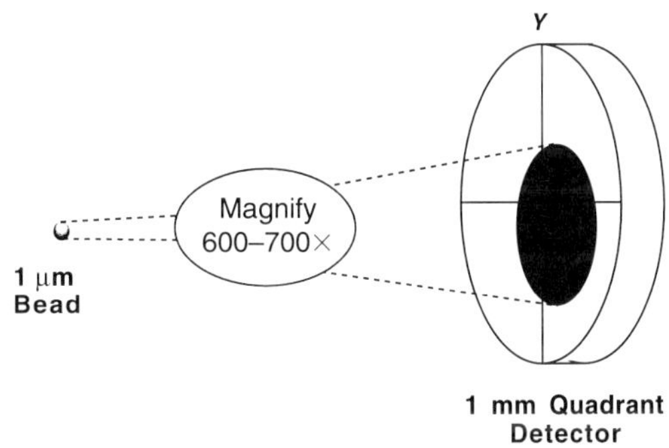

Fig. 3 Quadrant detector imaging. A bright-field image of a trapped bead is magnified 600- to 700-fold and then focused onto a four-element quadrant photodetector. Each quadrant converts incident light to current, which can then be analyzed to determine the position of the bead with resolution several orders of magnitude beyond the diffraction limit.

VI. Noise Sources

Position-detection resolution is limited ultimately by several sources of noise. The nature of these noise sources must be considered so that appropriate compromises can be made in instrument design to match experimental requirements. These noise sources include mechanical vibrations, statistical fluctuation in light intensity, and electronic noise. Vibrations are handled by minimizing external sounds, by mounting the apparatus on a vibration-isolated table, and sometimes by encasing the optical pathway. Electronic and light intensity noise are discussed here.

The intensity of the illumination source must be fairly stable because changes will be registered as anomalous bead movements. Even if the classical light intensity from the illumination source remains constant, as usually is the case for stable arc lamps or lasers, the light incident on each of the four quadrants will fluctuate randomly, a phenomenon known as shot noise. Given the quantum nature of light, the classical intensity simply reflects a probability density characterizing a distribution of arriving photons that will be statistical in nature. In other words, although a given number of photons will arrive in a fixed time on the average, the number arriving in any such unit of time will vary according to Poisson statistics. The net effect is to create a noise photocurrent of rms value:

$$<x^2>^{1/2} = (2eiB)^{1/2},$$

where i is the mean photocurrent determined by incident photon flux and quantum efficiency of conversion to photoelectrons, e is the charge of an electron, and B is the bandwidth over which the measurement is made. If the photocurrent is integrated for a longer time, or, equivalently, measurement bandwidth is reduced, less fluctuation will be seen in the signal. Because the signal varies with the light intensity and the noise varies with the square root of light intensity, a brighter light source will reduce shot noise relative to the signal level.

Detector resolution in the system described here is limited by shot noise. Given the arc lamp illumination, this photon fluctuation alone can account for a few nm of peak-to-peak position noise if the signal is examined with a 15-kHz bandwidth. Shot noise can be overwhelmed simply by boosting the light levels. As noted earlier, our instrument employs a 75-W xenon arc as the illumination source. More powerful lamps are available, but there is a fundamental limitation in the inability of incoherent illumination to provide an image brighter than the light source itself. Hence, the best light sources will have an optimal ratio of light power to source surface area, rather than simply an optimal output power. By this standard, stronger arcs tend to fare poorly because of increases in size.

A coherent light source, such as a laser, can provide a sufficiently bright image to overwhelm shot noise. However, this gives rise to a host of new problems. Whereas an incoherent imaging system is linear in light intensity, a coherent one is linear in complex amplitude. A corrugated specimen will introduce phase shifts erratically. Light scattered from the multiple object points contributing to an image point will appear as a phasor sum of randomly phase-shifted components. Thus, the image will appear granular and inhomogenous, a phenomenon known as "speckle," which plays havoc with a final image even if the light is only partially coherent. Moreover, because the spatial frequency response of a coherent optical imaging path has a very sharp cutoff, fringes appear around objects in the image. The beam can be focused to an approximately 10-μm waist and can illuminate the particle at the waist center. However, statistical correlations then appear between the detected x and y coordinates of a bead undergoing Brownian motion. These variables should be statistically independent, and appear as such in an incoherent image. The correlations are necessarily artifactual, and can be devastating in a closed-loop feedback circuit, which will be described below. Additionally, the image becomes extremely sensitive to vibrations, noise, or optical imperfections such as dust on lenses. Despite all of this, a laser beam, expanded so that the specimen falls well within a single bright fringe of the light diffraction pattern in the specimen plane, has been used to illuminate a similar particle in a comparable imaging system. However, even in this case, the detected position signals have not been stable at low frequencies ($<$1 Hz) (Ishijima *et al.,* 1996).

Another option involves scrambling the phase coherence of a laser light source. Progress in this direction has been modest. Future technologies may use optical amplifiers running outside of lasing cavities but with very powerful pump sources to amplify spontaneous noise and produce high-intensity, incoherent light. At the moment, such devices are expensive and available only in the 1.55-μm wavelength range, but no fundamental problems appear to face the extension of the method to a wider variety of amplifiers. We have employed a primitive version of this idea, a diode laser operated just at the lasing threshold with the beam passed through a multimode optical fiber with a mode scrambler, to create a bright and stable illumination source. However, we also detect occasional drift at low frequency.

Instead of creating a separate image of the trapping plane, the trapping light can be used directly to detect small displacements of the trapped particle. This can be done interferometrically (Ghislain *et al.,* 1994; Svoboda *et al.,* 1993) or simply by imaging the beam itself (Smith *et al.,* 1996). The trapped bead acts as a microlens, deflecting the trapping light as it moves. In fact, bead displacement along the trapping beam axis has been detected with 10-nm resolution by measuring the forward scatter of the trapping light (Ghislain *et al.,* 1994). In this situation, shot noise is overwhelmed by high levels of light signal intensity.

The shot noise will add in quadrature to two other variance sources, dark noise and Johnson noise, described later. Dark noise simply reflects the random excitation of photoelectrons even in the absence of light. This noise level is typically less than that of shot noise and becomes less important with increased light levels because the signal and shot noise will rise with light intensity, and the dark noise will not. Once again, one can overwhelm it by increasing image brightness, thus increasing the signal while the noise remains constant.

The tiny currents from the photodiode elements must be converted into larger, more stable voltage values in a preamplifier located physically close to the detector. Extreme care must be exercised in constructing the ideal current-to-voltage converters (I-V converters) for this purpose. These voltage signals can then be carried to separate circuits to be added and subtracted as needed to create signals proportional to bead displacement in both dimensions.

First, the feedback resistor must be fairly large. The aforementioned noise sources will add in quadrature to the Johnson noise voltage, which appears across the I-V converter feedback resistor and has rms value

$$(4k_{\mathrm{B}}TRB)^{1/2},$$

where $k_{\mathrm{B}}T$ is the thermal energy, R is the resistance, and B is the bandwidth over which the noise is measured. The signal is proportional to the light intensity, the radiant sensitivity of the detector, and the value of the feedback resistor R. Thus, the signal-to-noise ratio will vary as the root R necessitating a fairly large resistor in the I-V converter. These considerations argue against using a smaller R and amplifying the signal afterward. However, as in the preceeding cases, increased image brightness will augment the signal without affecting the noise, making this consideration less important.

Second, depending on the application, consideration may need to be given to the detection bandwidth limit imposed by the parasitic capacitance across the I-V converter terminals, illustrated in Fig. 4. A capacitor consists of any two separated conducing elements and will behave in predictable ways when voltage differences appear between the elements. In this case, the ends of the resistor are the plates forming a tiny, "parasitic" capacitance. The parallel combination of the feedback resistor and the parasitic capacitor will act as a low-pass filter, or integrator, with time constant set by the product of R and C. Ordinarily, this product is negligible. However, with very large resistor values of approximately 200M, the time constant (given a 7 pF capacitance, characteristic of our simply wired circuit) is about 1.4 ms, or, equivalently, the 3-dB corner frequency of the low-pass filter is 110 Hz. Two considerations now enter the picture. First, the detector is not fast enough to track the full extent of information beyond 110 Hz. In fact, the tracked amplitude has already dropped to 70% of the actual amplitude at 100 Hz. Second, information above 110 Hz will be shifted in phase. This phase lag is 45° at the corner frequency and drops sharply to 90° at higher frequency.

In some applications, neither consideration may be relevant. For instance, if noisy data must be filtered before examination, higher frequency information might be discarded even if it were available, and the parasitic capacitance is of little consequence. However, the Fourier transform spectrum of trapped particle motion will have a 3-dB corner frequency set by

$$f_{3Db} = \alpha/(2\pi b),$$

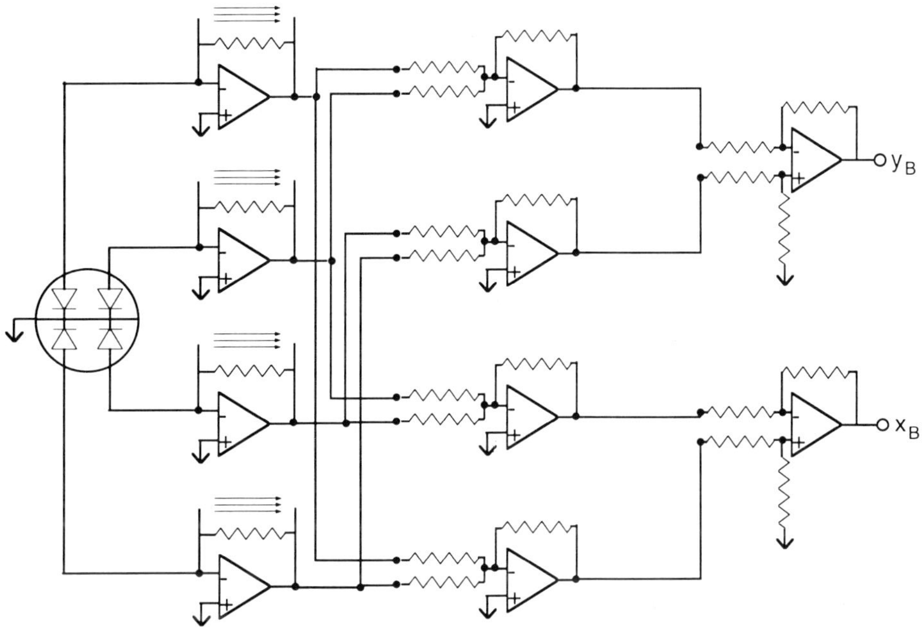

Fig. 4 Quadrant detector circuit. The detector itself is at the far left, with four independent photodiodes sharing a common ground. The first-stage I-V converters, immediately to the right of the detector, typically use very large feedback resistors in the range of 100 MΩ. The electric field lines shown above the resistors connect the exposed resistor terminals to create a parasitic capacitance element. OPA121 or OPA602 amplifiers are used because of low bias current levels. The second and third stages of the electronics, respectively, add and subtract the quadrant signals to generate x and y position readings. OP270 and AMP03 amplifiers are used in the second and third stages, respectively.

where b is the coefficient of viscous drag and α is the trap stiffness. This typically is considerably more than 110 Hz, so one must realize that the true extent of bead motion is not reflected in the data collected. Assuming that the trap stiffness has been measured independently, the true extent of motion can be computed based on the equipartition of energy. Every generalized coordinate, or degree of freedom describing the system, will have average energy of $1/2kT$, where k is Boltzmann's constant and T is the temperature in Kelvin. Thus, the rms value of bead position is

$$\langle x^2 \rangle^{1/2} = (kT/\alpha)^{1/2}$$

where α is the trap stiffness.

However, if the detector signal is used in a real-time control system, or analysis of the noise is important, the effect of the bandwidth limit must be considered. Moreover, if one is using detector signals in a closed-loop feedback circuit, the aforementioned phase lag can be particularly virulent, as discussed later.

If a circuit is simply wired on a typical circuit board, parasitic capacitances of approximately 7 pF can be expected. One way to minimize the problem is to reduce the value of the first stage feedback resistor R and hence the RC product. However, this comes at the cost of increased Johnson noise relative to signal, as discussed earlier. If the light levels are increased to further boost the signal, or particle motion is detected by imaging the trapping beam rather than the particle itself, this consideration fades in importance. A second way to address the issue involves minimizing the capacitances, typically by imprinting the circuit into a board to minimize lead lengths and by bending large resistors over extended ground planes. In principle, minimizing lead size will minimize the capacitance between leads. Moreover, many field lines from the resistor terminals should be intercepted by ground, thus creating harmless tiny capacitors between the terminals to ground at the expense of the dangerous capacitor across the terminals. Such circuits can be used to extend the bandwidth beyond 10 kHz, comfortably greater than the 3-dB frequency of bead motion for most trap stiffnesses. A third way to mitigate the problem is through electronic deconvolution of the position signals. This involves adding a weighted derivative of the signal back to the signal. The weighting coefficient must be set to match the integration time constant of the detector divided by the differentiator time constant. This method must be used with care because high frequency noise will be amplified as well. In fact, the gain must be rolled off to contain this. Typically, these parameters can be set by monitoring the detector response to abrupt changes in light. For instance, the trap itself can be imaged, small square-wave displacements can be introduced, and correction parameters can be set to minimize detector rise time. Practically speaking, given a 110-Hz bandwidth limit, such a circuit can extend the bandwidth to nearly 1 kHz, beyond which the signal is irretrievably lost in noise.

VII. Feedback

The effective stiffness of an optical trap can be greatly enhanced by means of a negative feedback circuit, as illustrated in Fig. 5. The bead displacement is monitored electronically and the trap position is adjusted to shift bead position to some desired value and maintain it once there. In this situation, displacement is monitored using the quadrant detector position signals, and control is implemented by driving the AOMs to move the trap.

A basic feedback scheme can be envisioned where, for any bead displacement X in a given direction, the trap is moved by some gain P times X in the opposite direction, a scheme known as proportional feedback gain. Although an unassisted trap will pull back a force of αX, where α is the trap stiffness, the feedback-enhanced trap now pulls with $\alpha(X + PX)$ and the effective trap stiffness has been increased by a factor of $P + 1$. Moreover, changing the effective trap strength also affects the time constants of the system. If an unassisted trap is

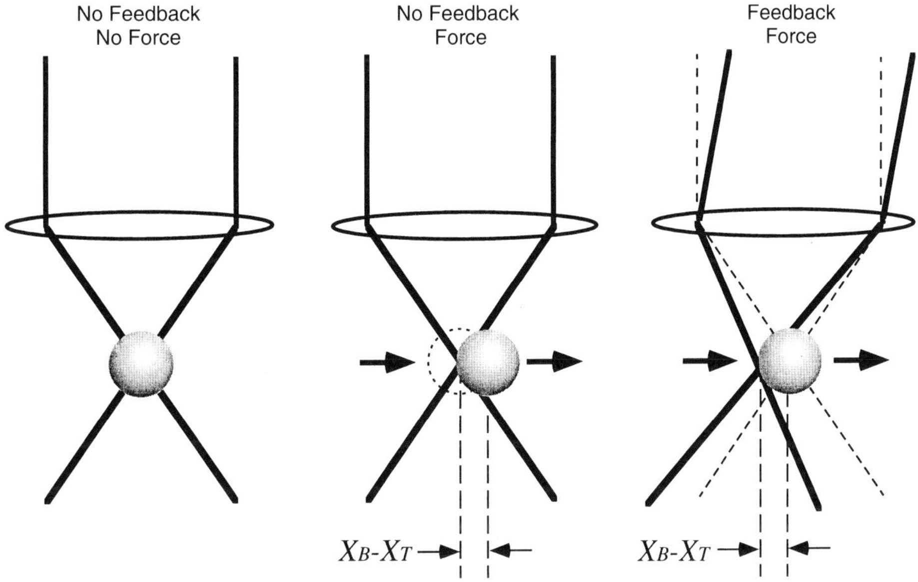

Fig. 5 Principle of feedback control. In the left extreme depiction, a bead is held by an optical trap in the absence of external forces. The bead will rest at the potential energy minimum, just below the laser focal point. In the center, an external load (heavy arrows) has been applied to the bead. The bead has moved until this load is matched by the restoring force exerted by the trap. At the right, an active feedback loop has moved the trap to the left to match the rightward force such that the bead moves very little. Note that the distance between the bead and trap center, and thus the force exerted by the trap, is identical in these situations. The only difference is whether this relative displacement is caused by motion of the bead or of the trap. X_B refers to bead position, and X_T refers to trap position.

displaced suddenly, the bead will move to the new trap position with a time constant of α/b, where b is the coefficient of viscous drag. Thus, as α is effectively increased, the system will have a faster response. However, the system will become less well damped and eventually underdamped, then unstable as P continues to rise. Thus, if the purpose is simply to create a stronger trap, proportional gain can often suffice. However, if the purpose is to eliminate as much bead motion as possible, additional gain types will be necessary.

Even in theory, proportional gain alone cannot pull the bead to a desired position. The feedback correction signal is proportional to the "error," or the deviation of the bead from this position. Once this error is reduced to zero, the correction signal will vanish. As mentioned earlier, the error cannot simply be made arbitrarily small by increasing the gain. This problem typically is addressed using integral control, in which an additional correction signal is proportional to the time integral of the error signal. Thus, the current correction signal can be the result of the past errors, in theory allowing the system to reduce the steady-state error to zero. Integral gain tends to slow down the transient response of the system, and can quickly become unstable as the gain is increased if it is used alone. Typically, it is used in conjunction with proportional gain. Integral control provides the highest gain at very low frequencies, meaning that slow deflections of the bead will trigger particularly large correction signals. In fact, the gain is typically rolled off at low frequency to prevent extremely high DC gain from saturating the circuit.

Integral gain will improve steady-state tracking, but there remains the problem of underdamping and instability as the proportional gain is increased to suppress high frequency motion. This problem must be addressed using differential gain, also called velocity gain, normally used in conjunction with proportional gain to increase the damping and improve the stability. The effect of differential gain in the bead equation of motion is simply to augment the coefficient of viscous drag. Thus, although proportional gain will reduce the time constants of the system, differential gain will increase them. This allows proportional gain to be increased well beyond the level at which it alone would cause instability. Moreover, differential gain output will increase with frequency and can be essential in countering very fast noise. Because the trap deflection will depend on the rate of change of bead position, the control is "anticipatory." Although pure derivative gain decreases high frequency noise, the bead position becomes unstable quickly if perturbed. Standard practice is to use all three gains in conjunction.

A combination of proportional, integral, and differential gain (PID) control allows one to simultaneously reduce the bead motion while maintaining stability, to control both steady state and transient deviations from the desired position. The gains typically are set just below the level that causes the bead to oscillate. This can require some finesse because increased derivative gain can allow the proportional gain to be increased still further before feedback instability is triggered. We have also found that the bead tends to escape the trap if all feedback gains become active simultaneously. Stability of the initial lock is increased when the integral gain becomes active after the proportional and derivative gains.

Such a feedback circuit can be implemented by using either analog or digital electronics. Analog integrators and differentiators are relatively simple to construct, but care must be taken to roll off the gains as mentioned earlier. A digital system requires specialized digital signal-processing electronics to rapidly sample and process the signal. Several design methods exist for conversion of analog filter parameters to digital filter coefficients. Although analog feedback seems to offer superior performance, digital feedback is more flexible. The circuit can be modified by changing a few lines of code in control system software rather than by incorporating additional components onto a circuit board.

At trap strengths typical of motor assays, an optically trapped bead can have peak-to-peak thermally driven movements of approximately 40–50 nm. Negative PID feedback can clamp bead position within about 1 nm or so (Molloy *et al.*, 1995; Simmons *et al.*, 1996). While a bead is under feedback lock, external forces will be countered by trap movements to minimize changes in bead position.

Thus, the trap displacement necessary to keep the bead tightly localized is a direct measurement of any external forces acting on the bead. In motor assays, feedback has been used to clamp an optically trapped bead and measure the force exerted by a single motor molecule on an attached filament (Finer *et al.*, 1994; Molloy *et al.*, 1995). Additionally, proportional feedback alone has been used to increase the load placed on motor molecules pulling against the trapping force (Finer *et al.*, 1995).

A few notes of caution must be added here. First, because feedback operates by deflecting the trap, it would become very difficult to monitor bead position by imaging the bead with trapping light. Second, feedback can never restrict bead fluctuations to a level less than the corresponding noise in position detection. The circuit will respond to apparent but false bead motions, just as it does to real ones, by moving the trap to counter them. Much of this noise might appear to vanish, because the bead is being moved to create this appearance. This makes it especially important to reduce shot noise levels if a tighter feedback lock is desired. Third, great care must be exercised in interpreting feedback performance with a bandwidth-limited detector. Feedback will respond to information beyond the detector rolloff corner frequency, because the information is attenuated but not eliminated. However, as noted earlier, this information is also shifted in phase. Another important detail is that the optically trapped bead will follow the trap position with high fidelity only within a certain bandwidth. The 3-dB corner frequency is set by the trap stiffness and solution viscosity; beyond this, the amplitude of bead motion will fall and the bead will lag behind the trap in phase.

Note that integral gain will provide a phase lag, and differential gain a phase lead. In fact, these shifts are necessary for the circuit to perform as described earlier. However, added phase shifts between the position and signal (detector) and between the actuator and variable (trap) will affect the system in predictable ways, making it difficult to increase the gain to optimize performance at higher frequencies. Feedback oscillations occur when the loop phase lag rises to 180°. In other words, the information to which the feedback circuit effectively responds is completely out of phase with the actual bead motion. In fact, a feedback oscillation can easily occur at a frequency beyond the 3-dB corner frequency of the detector. Because the detector amplitude response is poor at higher frequency, such an oscillation might appear to have only a marginal effect on system performance, although in actuality it serves to *increase* bead position noise outside detector bandwidth. Be aware that slight abnormalities seen in bead motion power spectra at very high frequencies may reflect very large peaks seen through the band-limited detector.

VIII. Calibration

To calibrate the system described earlier, the magnification at the detector plane is measured by moving the detector through an image of a grating with

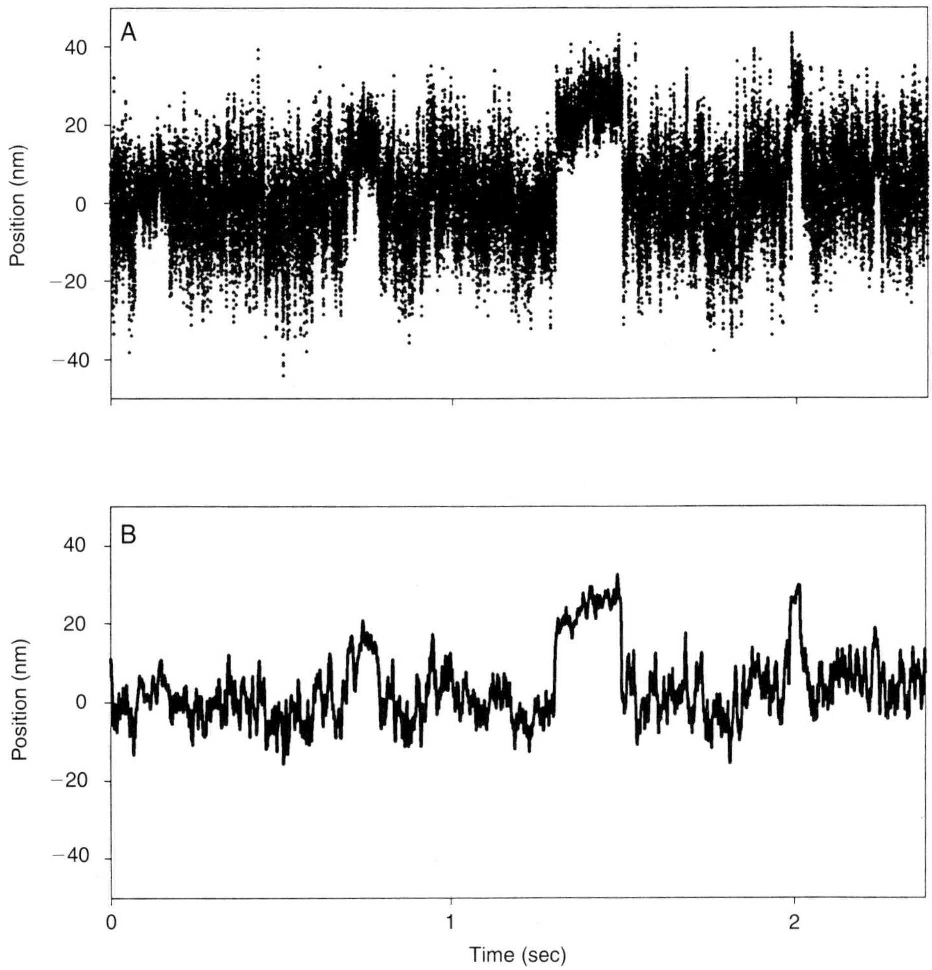

Fig. 6 Trace of bead position. Two optically trapped beads are attached chemically to a single actin filament, which is then moved into close proximity with surface-bound myosin molecules. (A) The raw data trace is shown. Transient deflections are caused by single molecules of myosin, which bind to the filament, arresting it somewhere in its range of thermal diffusion, and move it in a given direction defined by actin filament polarity. The proteins were in a 5-μM ATP solution. Trap stiffness is around 0.04 pN/nm; detector bandwidth is 15 kHz; and the sampling frequency is 20 kHz. The amplitude of thermal diffusion is significant relative to the biologically relevant distances we are attempting to measure, requiring statistical methods applied to large numbers of myosin-induced bead deflections. (B) The signal has been smoothed using a first-order Butterworth filter with a cutoff frequency of 100 Hz. This resembles a trace that could be seen with a detector of 100-Hz bandwidth. The faster components of thermal diffusion have been averaged, causing the motions to appear deceptively smooth. In many applications, the slower, underlying motions shown in B are of primary interest, and high detector bandwidth is not important. However, it must be remembered that traces such as those shown in B can often mask actual bead motion similar to that shown in *a*.

known dimensions. The deflection is then measured at the detector plane corresponding to a given AOM-induced bead deflection. Given the magnification, this can be converted to distance in the specimen plane. The position signal corresponding to this deflection is measured to determine a calibration constant in distance per volt. For beads 1 μm in diameter, the detector signal varies linearly with displacements less than 300 nm away from the trap center. This parameter, of course, will vary with magnification.

Calibration of trap stiffness can be done in four ways. First, one can apply solution flow and compute the force exerted on the bead via Stoke's law:

$$F = 6\pi\eta r v$$

where η is the coefficient of viscosity, r is the bead radius, and v is the solution flow speed. Second, one can measure the 3-dB corner frequency of the bead motion power spectrum and use it to compute the stiffness. Third, one can apply square-wave trap motions and measure the rise time with which the bead follows the trap. Note that the latter two methods assume a detector bandwidth that will encompass almost all of the bead motion. Moreover, they all depend on Stoke's law, which is perturbed by the proximity of a surface. Hence, the trap must be moved 5 to 10 μm deep into solution. Fourth, one can measure the extent of bead diffusion and use the equipartition theorem to estimate the stiffness. This also requires high detector bandwidth, but does not depend on Stoke's law and thus can be measured near a surface. Aforementioned force measurements require knowledge of trap stiffness because the product of the trap stiffness and the separation distance between trap and bead is the restoring force applied to the bead.

IX. Analysis

High-resolution position and force measurements as described tend to produce very noisy signals due to thermally driven Brownian motion as well as other noise sources. A sample trace of trapped bead position in an experiment measuring biological motor activity is shown in Fig. 6. In many applications, analysis of noise can be a useful way to extract meaningful data (Molloy *et al.*, 1995; Patlak 1993; Svoboda *et al.*, 1994). In other situations, one may seek to identify sharp transitions into and out of motor-bound states. In this case, the random noise can obscure the relevant data. Traditional signal-filtering techniques include local averaging with various windows or, equivalently, discarding high frequency information. Experiments can be done with various types of Fourier filters, including arbitrarily sharp-frequency cutoffs, with no phase distortion in the passband. Additionally, nonlinear, wavelet-based filters can be effective in removing high-frequency noise without corrupting excessively high-frequency components of data.

The Fourier basis is a natural one for infinite duration signals, but wavelet bases can be more relevant for the finite-duration data traces generated in these experiments. The basis is generated by dilation and shift of a "mother wavelet" function, of which a variety are available. Basis functions are thus localized in both time and frequency, including short high-frequency transients or longer low-frequency components. Although it is less straightforward to design, wavelet-based filtering can allow simultaneous preservation of sharp edges and removal of high-frequency noise. Reconstructing the time domain signal from a subset of the wavelet components often produces a better result than linear filtering in generating fairly clean and sharp data traces (Donoho 1993; Donoho 1995).

X. Conclusion

The optical trap has evolved into a useful tool for micromanipulation of cells and organelles as well as for precise, quantitative measures of tiny forces and displacements. The ultimate limit of the trap as a micromanipulator is probably the amount of force that can be generated by using a safe intensity of laser light. The weakness of light forces provides the sensitivity required in many applications, but precludes the high tension needed in others. In such situations, less compliant probes such as atomic force microscope cantilevers or glass micro-needles may be more appropriate.

The detection of small movements and forces is confronted only by the limits of detection, which presently involve the described technical hurdles that can be overcome. Beyond this, some measurements require means to constrain rotation of the trapped bead. Additionally, the force and displacement detection schemes discussed here apply not to the biological source but to the bead, which serves essentially as a probe. Future improvements should include reducing the compliance of probe–protein attachments and controlling the relative orientation of the interacting proteins. Moreover, more precise and reliable analysis to extract meaningful data from noisy signals will need to be developed to remove the observer bias from data tabulation. In this regard, we have shown that measurements of correlated thermal diffusion of optically trapped beads at either end of a single actin filament can be used to determine when a myosin molecule, which is attached to a nearby surface, binds to the actin (Mehta, *et al.,* 1997). Only solvable technical hurdles appear to challenge the increased precision and versatility required for the next generation of high-resolution mechanical measurements.

References

Ashkin, A., Dziedzic, J. M., Yamane, T. (1987). Optical trapping and manipulation of single cells using infrared laser beams. *Nature* **330,** 769–771.

Ashkin, A., Schutze, K., Dziedzic, J. M., Euteneuer, U., and Schliwa M. (1990). Force generation of organelle transport measured in vivo by an infrared laser trap. *Nature* **348,** 346–348.

Berns, M. W., Aist, J. R., Wright, W. H., and Liang, H. (1992). Optical trapping in animal and fungal cells using a tunable, near-infrared titanium-sapphire laser. *Exp. Cell Res.* **198,** 375–378.

Block, S. M. (1990). Optical tweezers: A new tool for biophysics. *In "Noninvasive Techniques in Cell Biology"* (J. K. Foskett and S. Grinstein, eds), New York: Wiley-Liss. *Mod. Rev. Cell Biol.* **9,** 375–402.

Donoho, D. L. (1993). Nonlinear wavelet methods for recovery of signals, densities, and spectra from indirect and noisy data. *Proc. Symp. Appl. Math.* **47,** 173–205.

Donoho, D. L. (1995). De-noising by Soft-thresholding. *IEEE Trans. Information Theory* **41,** 613–627.

Finer, J. T., Simmons, R. M., and Spudich, J. A. (1994). Single myosin molecule mechanics: Piconewton forces and nanometre steps. *Nature* **368,** 113–119.

Finer, J. T., Mehta, A. D., and Spudich, J. A. (1995). Characterization of single actin-myosin interactions. *Biophys. J.* **68,** 291s–297s.

Ghislain, L. P., Switz, N. A., and Webb, W. W. (1994). Measurement of small forces using an optical trap. *Rev. Sci. Instr.* **65,** 2762–2768.

Ishijima, A., Kojima, H., Higuchi, H., Harada, Y., Funatsu, T., and Yanagida, T. (1996). Multiple- and single-molecule analysis of the actomyosin motor by nanometer-piconewton manipulation with a microneedle: Unitary steps and forces. *Biophys. J.* **70,** 383–400.

Mehta, A. D., Finer, J. T., and Spudich, J. A. (1997). Detection of single-molecule interactions using correlated thermal diffusion. *Proc. Natl. Acad. Sci. U.S.A.* **94,** 7927–7931.

Molloy, J. E., Burns, J. E., Kendrick-Jones, J., Tregear, R. T., and White, D. C. S. (1995). Movement and force produced by a single myosin head. *Nature* **378,** 209–212.

Patlak, J. (1993). Measuring kinetics of complex single ion channel data using mean-variance histograms. *Biophys. J.* **65,** 29–42.

Perkins, T. T., Smith, D. E., and Chu, S. (1994). Direct observation of tube-like motion of a single polymer chain. *Science* **264,** 819–822.

Simmons, R. M., Finer, J. T., Chu, S., and Spudich, J. A. (1996). Quantitative measurements of force and displacement using an optical trap. *Biophys. J.* **70,** 1813–1822.

Smith, S. B., Cui, Y., and Bustamante, C. (1996). Overstretching B-DNA: The elastic response of individual double-stranded and single-stranded DNA molecules. *Science* **271,** 795–799.

Svoboda, K., Schmidt, C. F., Schnapp, B. J., and Block, S. M. (1993). Direct observation of kinesin stepping by optical trapping interferometry. *Nature* **365,** 721–727.

Svoboda, K., Mitra, P. P., and Block, S. M. (1994). Fluctuation analysis of motor protein movement and single enzyme kinetics. *Proc. Natl. Acad. Sci. USA* **91,** 11782–11786.

Visscher, K., Brakenhoff, G. J., and Krol, J. J. (1993). Micromanipulation by "multiple" optical traps created by a single fast scanning trap integrated with the bilateral confocal scanning laser microscope. *Cytometry* **14,** 105–114.

Wolenski, J. S., Cheney, R. E., Mooseker, M. S., and Forscher, P. (1995). In vitro motility of immunoadsorbed brain myosin-V using a limulus acrosomal process and optical tweezer-based assay. *J. Cell Sci.* **108,** 1489–1496.

CHAPTER 5

Laser Scissors and Tweezers

Michael W. Berns, Yona Tadir, Hong Liang, and Bruce Tromberg
Beckman Laser Institute and Medical Clinic
University of California at Irvine
Irvine, California 92612

I. Introduction

Current laser microscope techniques have their roots in the early work of the Russian, Tchakhotine, who from 1912 to 1938 published many studies describing the use of ultraviolet (UV) light focused into cells and eggs (Tchakhotine, 1912; see Berns, 1974 for a review of these studies). Soon after the advent of the laser in the early 1960s, Bessis in Paris described the first use of the laser to probe individual cells (Bessis *et al.,* 1962).

The early work of Tchakhotine and Bessis stimulated one of the authors (MWB) in 1966 to use a pulsed ruby laser microscope to study the development of a common millipede (Berns, 1968). For the past 30 years the authors have been engaged in developing and applying laser-based ablation ("scissors") microscope systems to the study of biological problems at the subcellular, cellular, tissue,

METHODS IN CELL BIOLOGY, VOL. 55
Copyright © 1998 by Academic Press. All rights of reproduction in any form reserved.
0091-679X/98 $25.00

and organism levels. Hundreds of studies have been published by the authors' group as well as by many others around the world. These studies have been reviewed periodically by various investigators (Berns, 1974; Berns and Rounds, 1970; Berns and Salet, 1972; Berns *et al.,* 1981; Berns *et al.,* 1991; Greulich and Leitz, 1994; Greulich and Wolfrum, 1989; Moreno *et al.,* 1969; Peterson and Berns, 1980; Weber and Greulich, 1992). Those series of reviews should be consulted for detailed descriptions of the technologies, the mechanisms of light interaction, and the biological problems studied.

After laser scissors, the next major advance in the area of laser-based subcellular manipulation was the development of optical trapping ("laser tweezers') by Arthur Ashkin (Ashkin, 1980; Ashkin and Dzeidzic, 1987; Ashkin and Dzeidzic, 1989; and Ashkin *et al.,* 1986; Ashkin *et al.,* 1987). The use of optically induced gradient forces permitted for the first time the noninvasive and nondestructive manipulation of organelles within the cell. These and other studies have been reviewed in Berns *et al.* (1991) and Greulich and Leitz (1994).

It is not surprising that these two noninvasive optical tools (laser scissors and tweezers) would be combined to provide the cell biologist with the capability both to hold (with tweezers) and to cut (with scissors) individual cells and organelles. The first combined use of these techniques was to induce cell fusion by first holding and positioning two cells with the laser tweezers, then cutting the adjoining cell membranes with the laser scissors (Fig. 1; Weigand-Steubing *et al.,* 1991). These combined techniques were subsequently applied to a variety of problems in chromosome manipulation, gene cloning, and egg–sperm interactions (see subsequent sections of this chapter for reviews of these studies). A versatile microscope workstation composed of two trapping laser beams and one ablation beam has been developed around a confocal fluorescent microscope, known as the Confocal Ablation Trapping System (CATS; Fig. 2).

Another major advance in microscope-based laser manipulation was the demonstration of two-photon-excited fluorescence by Watt Webb and his colleagues at Cornell (Denk, 1996; Denk *et al.,* 1990; Denk *et al.,* 1994). This technique permits focal plane specific fluorescence because the only point in the cell/organelle at which the photon intensity is high enough to result in two-photon absorption is the intense focal point of the laser beam. Thus the absorbing molecule absorbs two photons virtually simultaneously, and as a result behaves as if it has absorbed a single photon at one-half the wavelength of the impinging photons. The result is fluorescence at a wavelength shorter than the excitation wavelength.

Although two-photon fluorescence was developed initially as an analytical fluorescence technique, in the future it may be used as a technique to induce photochemical events at the cellular and subcellular levels. Preliminary work has demonstrated the production of UV (365 nm)-induced fluorescence of a psoralen molecule using the 730-nm beam from a mode-locked Ti : sapphire laser (Oh *et al.,* 1997). Although these studies have demonstrated two-photon-induced

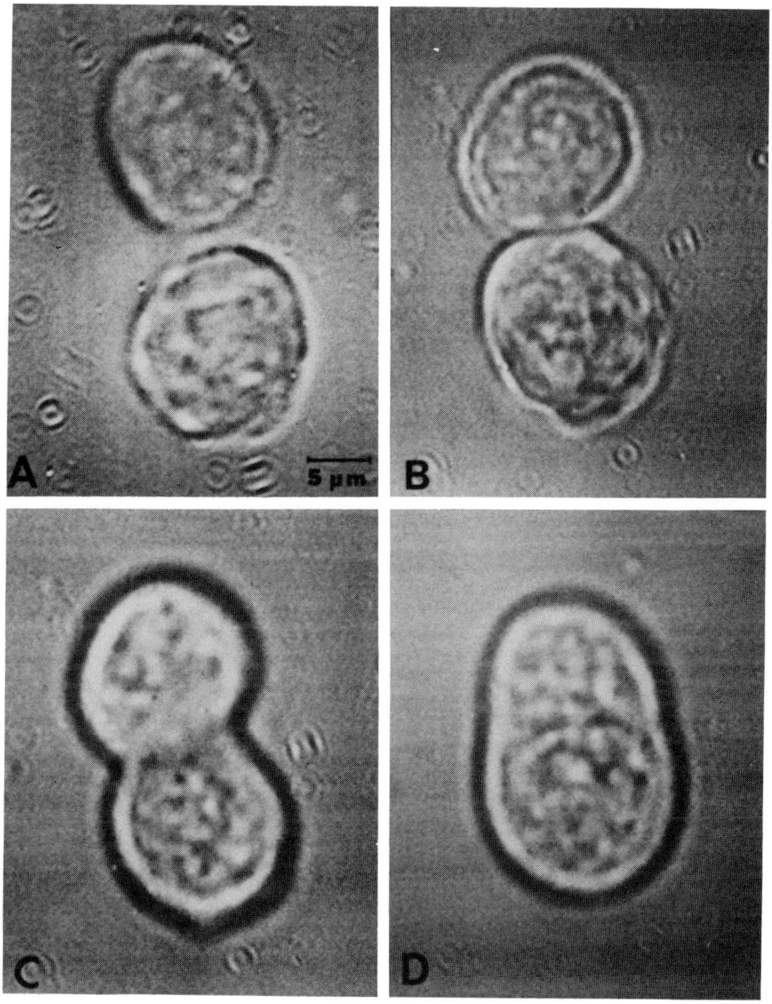

Fig. 1 Laser-induced cell fusion of one pair of cells in the optical trap. (A) The lower one of the two cells has been brought near the upper cell by dragging it using the optical trap. Note that the cells are not in contact yet. (B) Close cell contact is provided now after dragging the lower cell in the trap very close to the other cell. The UV laser beam was turned on at this moment. (C) After being hit by approximately 10 pulses of the 366-nm laser microbeam, the cells started to fuse. Note the separating plasma membrane disappearing. (D) About 5 min postfusion, the hybrid cell has rounded up.

fluorescence, it should be possible to produce two photon-induced molecular crosslinking within individual cells and at selected sites on targeted chromosomes. Multiphoton-induced focal plane specific photochemistry could become a powerful tool either to induce or to suppress site-specific photochemical processes in

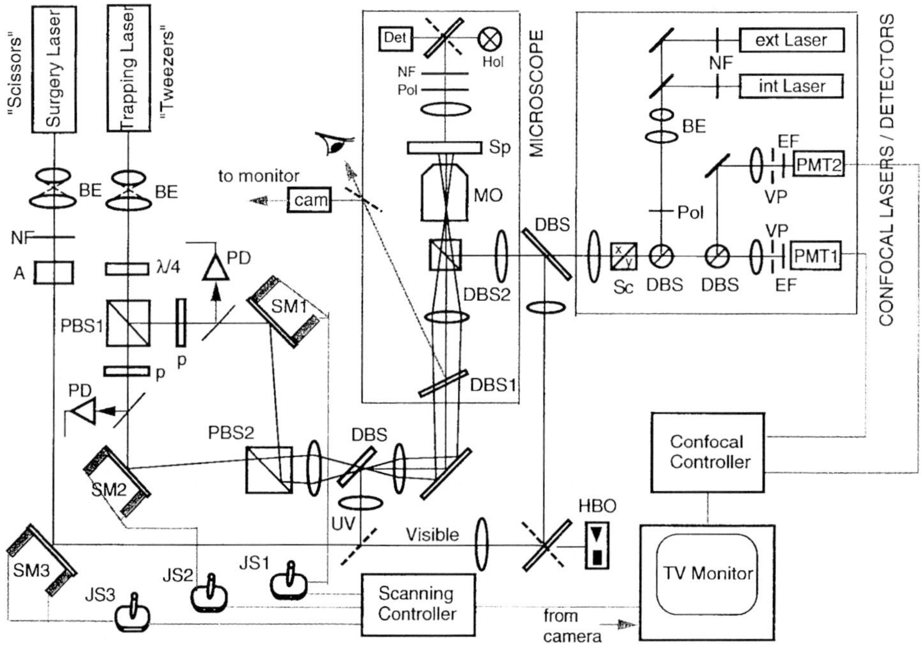

Fig. 2 Confocal Ablation Trapping System (CATS). Two external lasers are brought into the confocal microscope. The *tweezers* laser is a Ti:sapphire laser and is divided into two beams by prism beam splitter (PBS1) and recombined into two coaxial beams by PBS2. Movement of each of the two tweezer beams is controlled by scanning mirrors (SM1 and SM2), which are controlled by two joystick controllers (JS1 and JS2) The *scissors* laser is an ND:YAG laser that can operate at the fundamental 1.06 μm and the second (532 nm), third (355 nm), and fourth (266 nm) harmonic wavelengths. The scissors beam reflects off a joystick-controlled (JS3) scanning mirror (SM3) and enters the microscope by two different paths, depending on the wavelength used. Other elements in the diagram are as follows: BE, beam expander; A, attenuator; PD, photodiode detector; DBS, dichroic beam splitter; UV, ultraviolet reflector; CAM, video camera; MO, microscope objective; Sp, specimen; Pol, polarizer; NF, neutral density filter; Sc, x-y scanner; PMT, photomultiplier tube; HBO, mercury lamp.

single cells. This technique could be used alone or in combination with laser scissors, laser tweezers, or both.

II. Mechanisms of Interaction

A. Laser Scissors

The first laser microscope system employed a pulsed ruby laser at 694.3 nm with a pulse duration of 500 μsec. The energy in the 2- 5-μm spot was approximately 100 μJ (Bessis *et al.*, 1992; Saks *et al.*, 1965). This level of energy was probably not high enough to produce an effect such as multiphoton absorption or nonlinear optical breakdown of the cells. As a result, thermal effects were obtained by relying on naturally absorbing molecules (e.g., hemoglobin in red blood cells; Bessis *et al.,* 1962) or a vital stain (e.g., toluidine blue, which was applied to rabbit embryo cells before irradiation; Daniel and Takahashi, 1965).

Not long after these studies, the use of the pulsed (50 μsec) argon ion laser at 514 and 488 nm in combination with the vital stain, acridine orange, to selectively alter chromosomes was described (Berns *et al.,* 1969). The laser was focused to a 1-μm diameter spot and a 1.5-W peak power per pulse as it entered the microscope. The selective effects in the focused spot in these and the earlier studies with the ruby laser were likely caused by heating of the chromophore, which, in turn, caused thermal destruction of the structure/cell to which it was bound. However, at the threshold levels of irradiation with the argon laser on acridine orange vitally stained chromosomes, light-activated photochemistry may have occurred. Acridine orange is known to be a photodynamic molecule that can be excited to its triplet state, which can subsequently interact with oxygen-producing singlet oxygen.

The next major advance in laser inactivation of cells and organelles was the advent of the solid-state Q-switched neodymium:yttrium aluminum garnet (Nd:YAG) laser. This laser produces wavelengths at its fundamental wavelength of 1.04 μm, or at harmonically generated wavelengths at 532 nm (green), 355 nm (UV), and 266 nm (UV). In addition, this laser can be used to pump a dye laser, which provides tunable wavelengths throughout the visible and UV spectrum (Berns *et al.,* 1981). Compared to previous laser systems, these lasers produce beams with pulses in the nanosecond and picosecond ranges, with hundreds of millijoules per pulse. This translates into megawatts and gigawatts per square centimeter in a 1-μm diameter spot. Under these irradiation conditions, selective subcellular damage could be produced without the use of any applied or known natural absorbing chromophore. Many of the subcellular laser scissors effects are likely due to nonlinear multiphoton absorption leading to multiphoton-induced UV-like photochemistry or optical breakdown due to the generation of a microplasma with a high electric field and consequent acousto-mechanical effects. The first paper describing the possibility of multiphoton effects of a laser focused into a live cell through a microscope was published in 1976 (Berns, 1976) and subsequently confirmed in 1983 (Calmettes and Berns, 1983).

Thus, by the mid 1980s laser microscope systems were available that could be used selectively to alter cells and subcellular regions through the use of chromophore-activated absorption, or nonlinear multiphoton absorption. The technique employing exogenous chromopohores has been extensively applied to dissect the mitotic apparatus for the purpose of elucidating the role of the polar region and the centrosome in cell division (Berns *et al.,* 1977), as well as to elucidate the role of the centrosome in the control of cell migration (Koonce *et al.,* 1984). Naturally occurring chromophores, such as the respiratory molecules in the mitochondria, have been used to study individual cardiac cell contractility.

Nonlinear multiphoton absorption has been used to deactivate unstained kinetochore regions of the chromosomes in order to study the role of this organelle in chromosome movement, as well as to probe the cell division process in a fungus (these studies will be described in a subsequent section of this chapter). Multiphoton absorption also has been used to cut the chromosome and study the movements of the remnants with respect to mitotic spindle dynamics (Liang *et al.,* 1994). Finally, multiphoton-induced effects have been used to inactivate specific genetic regions on chromosomes in order to elucidate the role of the genes in cell function (Berns *et al.,* unpublished).

In their review Greulich and Leitz (1994) discussed the mechanism of subcellular microprocessing using a short (2–3 nsec pulsed nitrogen laser operating at 337 nm. They described a situation in which 1 μJ of energy is focused to a 1-μm spot on a subcellular target. In this situation they estimated that 0.1–1% of the UV energy is absorbed. At first glance, it seems that the cellular material would be heated up to 1000 to 10,000 degrees centigrade, thus resulting in total evaporation of the target material, which, in turn, would change the absorption coefficient so that virtually all of the remaining incident energy would be absorbed. In this situation the local temperature rise would be millions of degrees, and a microplasma would form. Under these conditions the subcellular target as well as the entire cell would be destroyed. However, in reality, this does not happen. In fact, very precise subcellular microsurgery can be attained using these laser parameters. As pointed out in their article (Greulich and Leitz, 1994) the temperature rise is localized to a very small region of the cell because after 10 nsec the heat has dissipated into the surrounding environment (which is mostly water). They theorized that because of this, the temperature rise is quite low (probably only a few degrees). They concluded, that owing to the low temperature rise of the surrounding cell environment, the laser pulse is highly destructive only at the focal point on the target. The morphological observations of subcellular damage would support this theory. Furthermore, the same concept should hold for two-photon absorption of the 532-nm second harmonic Nd:YAG laser. The target would actually "see" two photons of 532 nm, which would be equivalent to one photon at 266 nm. The absorbed two quanta could generate either UV photochemistry, or thermal effects similar to those described by Greulich and Leitz for the 337-nm beam.

However, the situation may not always be as clear-cut as just described. For example, we frequently use one or two pulses of the 10-sec 532-nm frequency-doubled Nd:YAG laser beam to produce small lesions on unstained chromosomes in live cells (Fig. 3). These lesions appear as a change in refractive index on the chromosome as viewed through the phase-contrast microscope. By employing transmission electron microscopy, these lesions were shown to be structurally contained within the actual focal point of the laser (Rattner and Berns, 1974) with the surrounding region of the chromosome and the cytoplasm unaffected. This is precisely the situation described by Greulich and Leitz (1994) except that we used a 532-nm beam instead of a 337-nm beam. We have theorized that this

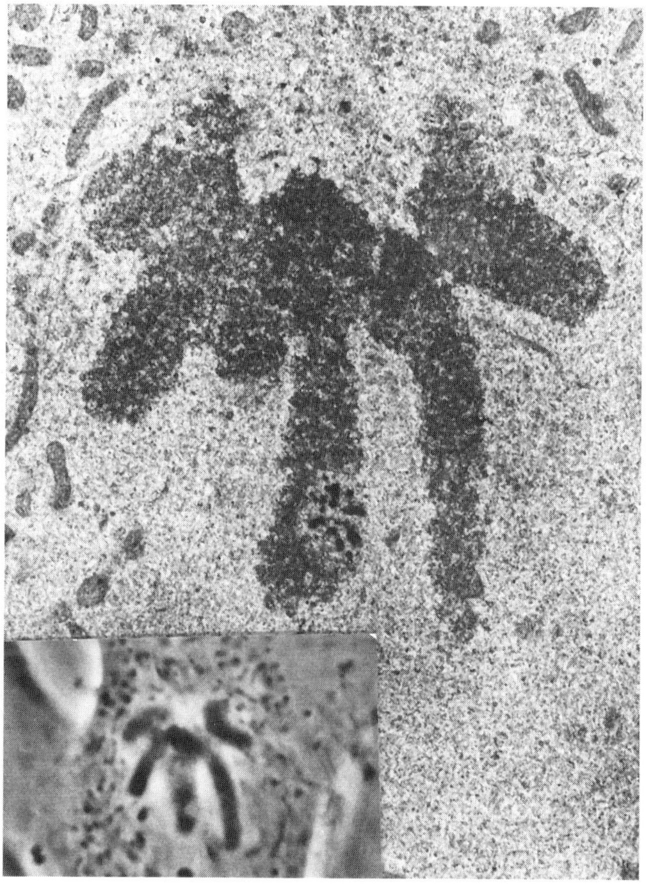

Fig. 3 Chromosome surgery. Insert is a phase-contrast live image of a dividing rat kangaroo kidney cell that has been exposed to a focused pulsed laser beam (magnification = 2,000×). Note the small lightened spot on the chromosome arm that indicates the selective damage produced by the laser beam. The larger picture is a transmission electron micrograph (magnification = 12,000×). The light-appearing damage in the phase-contrast image can be seen to be precisely confined to the chromsome in the electron micrograph (from Rattner and Berns, 1974).

type of affect is caused by multiphoton absorption (Calmettes and Berns, 1983). However, occasionally (less than 10% of the time) one 10-nsec laser pulse in the nJ–μJ regime focused to a 1-μm-diameter spot on a chromosome will result in an explosive event that not only destroys the cell completely, but also chips the glass surface to which the cell is adhered and destroys several of the surrounding cells by the creation of an acoustomechanical shock wave. Clearly in this situation a microplasma has been generated. The only variable in these experiments is the 10% variation of laser output from pulse to pulse for the Nd:YAG lasers. The chemical composition of the target (chromosome) and the size of the focal spot were not changed. The possible explanation for this observation is that the threshold of the transition from multiphoton-induced photochemistry and/or heat as described by Greulich and Leitz (1994) to the generation of a microplasma is within the 10% variation of the laser output. Thus, at the upper end of the 10% variation, the concentration of photons in the 1-μm-focused spot may be high enough for plasma generation. Regardless of whether or not this is the explanation for the observations, the fact remains that the mechanisms of subcellular photodisruption in a 1 μm-size spot are complex and not yet fully understood.

A final mechanism of laser interaction that should be mentioned involves two-photon-induced photochemistry of an applied chromophore. As mentioned previously, preliminary work has demonstrated the production of UV (365-nm)-induced fluorescence of a psoralen molecule using the 730-nm beam from a mode-locked Ti:sapphire laser (Oh *et al.,* 1997). Although these studies have demonstrated two-photon-induced fluorescence, it should be possible to produce UV-induced molecular crosslinking within individual cells at selected sites on targeted organelles. This technique could be applied generally within multicellular systems as well as within different target regions of a single cell.

B. Laser Tweezers

In addition to light quanta (photons) that ultimately may be used to alter a target (as discussed previously), light can be shown to have momentum that can be imparted to a target. The change in momentum over time produces a force on the object: $F = \alpha \times P/c$, where α is a dimensionless parameter between 1 and 2, P is the laser power, and c is the speed of light. This concept can best be illustrated using a ray optics diagram in which the object to be "trapped" is relatively transparent to the trapping beam and spherical. The transparent property is essential because the light beam will pass through the object without significant energy being absorbed by the target and either will be converted to heat or will generate photochemistry that may damage the target as discussed in the previous section. The curved geometry of the target is important because as the photons are refracted by the curved surface, the direction and amount of net force is affected. The direction of the forces on the object are a function of curvature, size, and angle at which the incident beam strikes the target (Fig. 4); the forces may be axial (toward or away from the source of the laser) or transverse (horizontal to the plane of incidence). The magnitude of the forces that can be applied to a biologic objects using a 25–300 mW 1-μm-diameter focused laser beam is in the range of piconewtons and is more than enough force to trap and move cell organelles as well as whole cells.

One key question raised concerns the degree of heating that optical traps may cause in the trapped object. Because a most frequently used optical trap is the 1.06-μm beam from the Nd:YAG laser, a study was undertaken to determine the magnitude of temperature increase. Temperature rise has been demonstrated through the use of an *in vitro* microfluorometric technique (Liu *et al.,* 1994). In

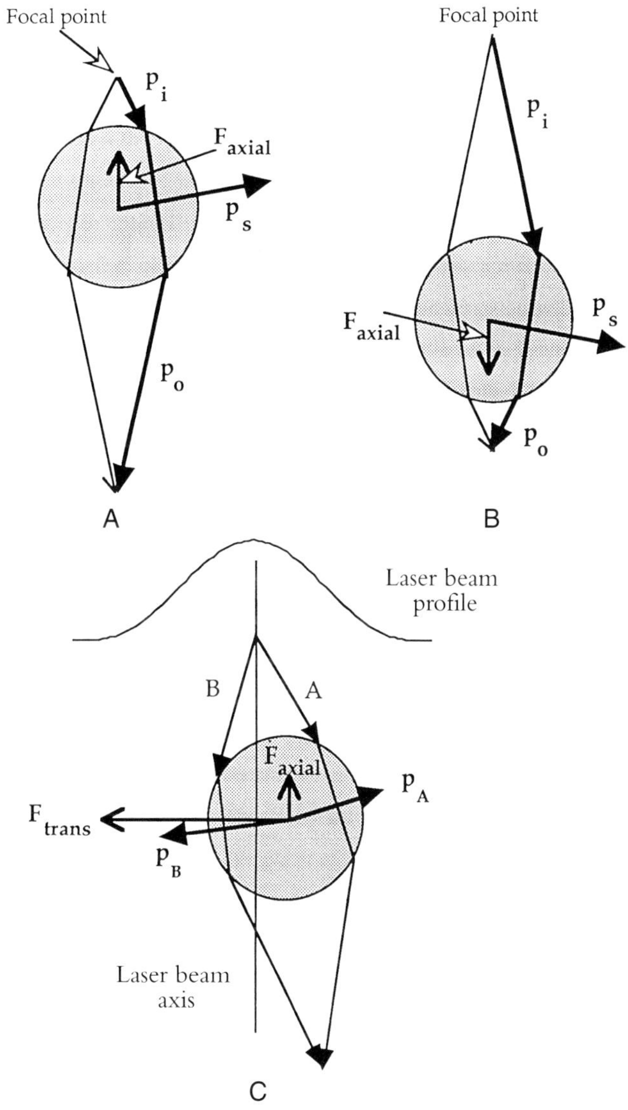

Fig. 4 Ray optics description of optical trapping. (A) Resultant axial force on a sphere when the convergence angle of the rays exiting the sphere is less than the angle of the rays entering the sphere. The resultant force is directed toward the beam waist. (B) When the convergence angle of the rays exiting the sphere is increased, the resultant force is directed away from the beam waist, and the sphere is not trapped. (C) Transverse restoring forces are directed toward the beam axis along the direction of momentum change on the sphere. (Courtesy of W. H. Wright, Ph.D.).

these studies, liposome membranes were labeled with the UV-absorbing dye, Lourdan. This dye undergoes a phase change with an increase in temperature that results in a red shift of the fluorescence emission spectrum as well as a decrease in fluorescence intensity. Liposomes 10-μm in diameter held in a 1.06-μm optical trap from a Nd:YAG laser exhibited a 1.1°C increase in temperature/100 mW.

The early laser tweezer systems employed the continuous-wave visible blue green argon ion lasers followed by the infrared Nd:YAG lasers (Ashkin and Dziedzic, 1987). Although the argon laser could be used for trapping, the absorption of the light by natural chromophores in the cells resulted in heat-induced cell damage. The lower absorbency of cell molecules to the 1.06-μm YAG laser made this the wavelength of choice for optical trapping. As discussed earlier, infrared (IR) traps of 40–300 mW could produce a localized temperature rise inside a cell of 1°C/100 mW. Thus, the most that could be expected in a cell under normal trapping conditions would be a temperature rise of 1–3°C. In experiments for which temperature stability is essential (such as in an enzyme-driven process), this may affect the cell process being studied. In most other situations, however, the temperature rise from the IR trap should be negligible as long as the power in the optical trap is below 300 mW.

A study of a range of trapping wavelengths was possible with the advent of the argon ion laser-pumped Ti:sapphire laser (Vorobjev *et al.*, 1993). This system provided near IR wavelengths from 700 to 900 nm. In an initial study, metaphase mitotic chromosomes were held in the laser trap with 130 mW from 0.3 sec to 5 min, then released to see if the cell could continue through a normal mitosis. Distinct adverse effects such as chromosome bridges were observed when cells were exposed to the trap at 760–765 nm. Minimal effects were seen at 700 nm and 800–820 nm. At 840 nm, damage effects were beginning to be detected, again with increasing frequency (Fig. 5).

In a more recent study (Liang *et al.*, 1996), cell cloning was assayed after exposure to different optical trapping wavelengths. Trapping powers in the focal point of 88 and 176 mW were used for time periods of 3–5 min. As in the previous study, wavelength-specific effects were detected. The highest cloning efficiencies were observed at 800, 990, and 995 nm. The worst wavelength was 760 nm. At 88 mW at a wavelength of 1.06 μm, the cloning efficiency was 60% after 3 min of trapping and less than 10% after 10 min of trapping. This compares to cloning efficiency of more than 80% at 990 nm for the same trapping duration. On the basis of this and the previous studies, it is concluded that care should be taken in choosing the appropriate trapping wavelength, and 760 nm should be avoided.

The mechanisms of the trap-induced cell damage are not known. On the basis of previously described thermal studies, it seems unlikely that heat damage was sufficient at the wavelengths and powers used. Multiphoton effects are a distinct possibility.

Two studies have reported two-photon-excited fluorescence in cells during trapping (Konig *et al.*, 1995; Liu *et al.*, 1995). In the first study, a 1.06-μm IR

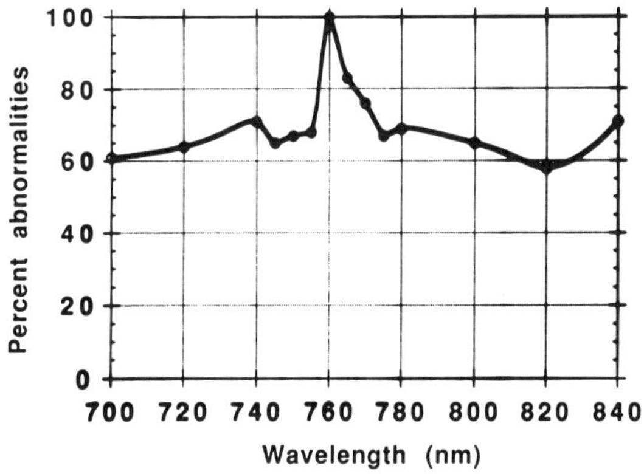

Fig. 5 Plot of the percentage of abnormal mitoses induced with different wavelengths under constant power of 130 mW. All cells were exposed for 0.3 sec to 5 min. The beam was focused on the clearly visible shoulder of a large chromosome. Abnormal mitoses were classified as either chromosome bridges during anaphase, or chromosomes failing to separate.

beam was focused as a trap into either sperm or Chinese hamster ovary cells that had been stained with the fluorescent dye, propidium iodide. The results clearly demonstrated fluorescence emission at a shorter wavelength than the excitation wavelength (1.06 μm). The two-photon-excitation wavelength of 530 nm matched the peak absorption wavelength of the propidium iodide. Furthermore, when fluorescence intensity was plotted as a function of laser trapping power, the intensity varied with the square of the incident laser power. This is a strong indication of a two-photon-driven process.

In the second study, 70 mW laser tweezers at 760 and 800 nm were compared with respect to cell cloning and two-photon-excited fluorescence in sperm cells (Konig *et al.,* 1995). The results demonstrated two-photon fluorescence in both situations. The fluorescence probe used, propidium iodide, was a live/dead probe that changed fluorescence color from green (probe bound to the membrane of a live cell) to red (dye accumulates in nucleus of a dead cell). When the 760-nm trap was used, the probe turned red within 62 sec of trapping, whereas with the 800-nm trap, the cells exhibited viable green fluorescence for more than 10 min while in the trap. These results compare favorably with the cell cloning studies that demonstrated a major loss of cloning efficiency in a 760-nm trap as opposed to a 800-nm trap.

In a third study (Konig *et al.,* 1996), damage was observed when cells were exposed to a near infrared multimode optical trap.

III. Biological Studies

A. Chromosome Surgery/Genetics

The early work using the red (694.3-nm) ruby laser (Bessis *et al.,* 1962; Saks *et al.,* 1965) focused primarily on damaging whole cells as opposed to small subcellular regions and organelles. With the advent of the shorter-wavelength blue-green argon ion laser, it was possible to expose subcellular regions and organelles to wavelengths of light that were (a) focused to smaller spot sizes because the wavelength was shorter and the mode structure of the beam was in the TEM00 mode (Berns 1974; Berns *et al.,* 1981); (b) absorbed by the target region by use of applied chromophores, such as acridine orange, which binds to the chromosomes (Berns *et al.,* 1969); and (c) absorbed by the target by natural chromophores such as the cytochromes in the mitochondria (Berns *et al.,* 1970a, Salet *et al.,* 1979).

The early studies on selective laser irradiation of chromosomes in dividing tissue culture cells demonstrated that a submicron pretargeted region of a chromosome could be destroyed without structurally affecting the rest of the chromosome or the cell (see Fig. 3). With the advent of more powerful short-pulsed lasers, it was possible to perform this type of subcellular microsurgery without using an applied chromophore; this effect was attributed to multiphoton absorption (Calmettes and Berns, 1983).

The genetic application of this technical capability has been in two areas: (*a*) inactivation of the ribosomal (nuclear organizer) genes (Berns *et al.,* 1979; Berns *et al.,* 1981) and (*b*) chromosome microdissection followed by gene cloning using PCR techniques (Djabali *et al.,* 1991; He *et al.,* 1997; Monajembashi *et al.,* 1986).

Nucleolar gene inactivation was first demonstrated in primary tissue cultures of salamander lung cells. These cells are very flat; the chromosomes are large; and there are three distinct nucleolar organizer chromosome regions clearly visible as secondary constrictions at the tips of the chromosomes. Thus, it was possible to laser irradiate one or more of these regions and demonstrate a concomitant loss in nucleolar activity in the postmitotic daughter cells (Berns *et al.,* 1970b). However, because these cells were in primary tissue culture, it was not possible to isolate and clone them to determine if the laser-induced gene alteration was maintained as a deficiency in subsequent cell generations.

The heritability of the laser-induced genetic deficiency was demonstrated by irradiating the nucleolar organizer region of dividing cells from the rat kangaroo, *Potorous tridactylis* (PTK2). Like the salamander lung cells, these cells remain flat throughout mitosis, and the chromosomes have clear secondary constrictions (nucleolar organizers) on their chromosomes. It was possible to clone populations from single cells that had one nucleolar organizer region inactivated by the laser microbeam (Berns *et al.,* 1979). All the descendant cells were deficient in one nucleolar organizer region and one group of ribosomal genes (rDNA). Using *in situ* molecular hybridization, it was possible to demonstrate that the rDNA was absent from the chromosome that was the "clonal" descendant from the irradiated chromosome (Berns *et al.,* 1981). This series of studies, which spanned 10 years, not only demonstrated the technical feasibility of "directed" genetic microsurgery, but also provided a better understanding of the function and regulation of the ribosomal genes.

The second area of laser-directed genetic analysis has been in the area of gene cloning and polymerase chain reaction (PCR). With the demonstrated technical ability to cut small regions of chromosomes without affecting the adjacent chromosome regions, it was possible to combine this technique with PCR (Djabali *et al.,* 1991; He *et al.,* 1997; Monajembashi *et al.,* 1986). There has been a need for a technique to replace tedious microneedle chromosome dissection with a more user-friendly, accurate, and time-efficient process. Although much DNA sequencing after gene amplification can be done on DNA fragments obtained by enzymatic digestion of chromosomes, there are a significant number of genes that can be located and cloned only after mechanical microdissection of a small submicron region of the chromosome of interest. In these studies the chromosomes have to be prepared as isolated chromosome spreads or suspensions. Then needles are used to dissect and pick up the desired chromosome region, which is subsequently subjected to PCR and sequence analysis.

Laser scissors have been developed instead of the microneedles to perform the chromosome microsurgery. This approach is being used to isolate and clone a region on human chromosome 9, which is suspected of containing one of the major genes for the disease, tuberous sclerosis (He, 1995; He *et al.,* 1997). The laser scissors technique was compared to the standard microneedle dissection technique. The results demonstrated that the PCR DNA insert size from the laser-dissected chromosomes averaged 450 base pairs, as compared to 250 base pairs for the microneedle-dissected chromosomes. Because the larger insert size is desirable for gene sequencing, it was concluded that laser microdissection is superior to the microneedle method in terms of insert size as well as with respect to ease of operation and speed of performing the procedure.

B. Mitosis and Motility

One of the largest series of laser scissors/tweezers studies has been in the area of cell mitosis and motility. This is because (*a*) laser microbeam technology has been particularly suited for probing these aspects of cell biology; (*b*) the cellular preparations are ideal for cytological manipulation (the chromosomes and spindles of flat cells are readily visible under light microscopy); and (*c*) the early work of Zirkle (1957, 1970) established the application of ultraviolet microbeams to cell motility problems.

Our studies have focused on (*a*) the mitotic apparatus (the pole, chromosomes, and spindle fibers), and (*b*) the cytoskeleton (microtubules and centrosome).

The mitotic studies have been conducted in rat kangaroo kidney cells (*Potorous tridactylis,* PTK2), primary cultures of the salamander (*Taricha granulosa*), and the fungus, *Nectria haematococca.* The precise methods used in the PTK2 studies are described in detail elsewhere (Berns *et al.,* 1994).

The early mitotic studies in the animal cells involved selective deactivation of the centrosome and the kinetochore region of the chromosome in an attempt to elucidate the functions of those regions in chromosome cell division (see summary of results in Berns *et al.,* 1981). In the centrosome studies, different applied and naturally absorbing molecules were used in combination with different laser wavelengths to functionally dissect the centrosome region, which is comprised of two major structural elements (the centrioles and the pericentriolar material). By inactivating one or the other of these structures, it was possible to demonstrate that the centriole was not necessary for completion of cell division once it had started, and that the pericentriolar material most likely was the microtubule organizing center. However, it was never determined if the cell could undergo a subsequent cell division once the centrioles had been destroyed.

Studies that involved laser inactivation of the kinetochore region of the chromosome revealed that poleward forces are exerted on the chromosomes long before they line up at the metaphase plate. When the kinetochore was destroyed on one side of a double-chromatid chromosome in late prometaphase, both attached chromatids rapidly moved toward and through the metaphase plate toward the pole that the unirradiated chromatid was facing. In addition, when the rest of the mitotic chromosomes underwent separation and anaphase movement toward the poles, the two chromatids that had moved prematurely toward one pole also separated from each other. The chromatid with the destroyed chromatid did not proceed toward any pole. This result demonstrated that the initial separation of chromatids in anaphase is not a force-mediated event (McNeill and Berns, 1981).

More recent studies have employed both the laser scissors and tweezers to study chromosome movements. The first laser tweezers study on chromosomes demonstrated that free chromosomes in suspension (outside of cells) could be easily moved about, and that chromosomes on the mitotic spindle could be manipulated (Berns *et al.,* 1989). However, the later studies were somewhat unusual in that the chromosomes seemed to be "pushed" by the tweezers, rather than pulled. These observation have never been adequately explained.

In later studies, it was possible by use of the tweezers to move whole chromosomes, arms of whole chromosomes, and chromosome fragments in dividing mitotic cells (Liang *et al.,* 1993; Liang *et al.,* 1994). In one of these studies (Liang *et al.,* 1994), a pulsed 532-nm second harmonic Nd : YAG laser scissors was used to cut the salamander chromosomes, and a CW 1.06-μm Nd : YAG laser was used to move the chromosome fragment in the cell. By employing known viscosities of the cytoplasm and recording the speed by which the chromosomes could be moved by the tweezers, it was possible to calculate that the forces generated by the tweezers necessary to move the chromosomes were in the range of 26–35 pN. This compared favorably to other studies that had shown that 1–74 pN was produced by nascent microtubles in the same cells (Alexander and Reider, 1991).

The preceding study and an earlier study employing a combined UV (366-nm) scissors and 1.06-μm tweezers to induce cell fusion (Wiegand-Steubing, 1991) were among the first studies employing both the laser scissors and tweezers to study cells.

Studies on mitosis in the fungus have provided major new insights on the mechanism of cell division in this organism. Early studies demonstrated that using the laser scissors to sever the central band of microtubules in the mitotic cell resulted in the chromosome movements toward the poles actually speeding up rather than decreasing. This suggested that the central spindle microtubules were actually functioning as a rate "governor" to hold the chromosomes back in response to astral forces pulling from the poles (Aist and Berns, 1981). Later studies demonstrated that laser inactivation of one pole resulted in the rapid movement of the chromosomes to the other pole. This result demonstrated the existence of "astral pulling forces" in the fungus. Because fungi do not have centrioles at their poles, the existence of forces emanating from noncentriolar material further raised the question concerning the role of the centriole in animal cells.

One possible role of the centriole in animal cells is to control the rate and direction of cell migration through interaction with the cytoskeleton. This idea was suggested by Albrecht-Beuhler and Bushnell (1979) who demonstrated a distinct orientation of the centriole with respect to the direction of cell migration. This theory was tested experimentally by using laser scissors to destroy the centriole in newt white blood cells (eosinophils) that were migrating in a straight line. In combination with a computer-based cell-tracking system (Berns and Berns, 1982), it was possible to demonstrate that cells with destroyed centrioles lost all ability to migrate in a straight direction, and that they moved with much reduced velocity (Koonce *et al.,* 1984). On the basis of this study and the previous studies in mitotic cells, it is suggested that a major role of the centriole is the control of cell migration.

C. Membrane Studies: Optoporation and Cell Fusion

In addition to manipulation of organelles within the cell, laser scissors also can be used to manipulate the outer cell membrane. The first extensive use of the laser on the cell membrane involved selective photobleaching of membrane-bound fluorescent probes (Koppel *et al.,* 1976). In these studies, a UV- or blue-wavelength laser was focused to a small micron-size spot on the cell surface in order to "bleach" the membrane-bound probe in that region. By monitoring the recovery of fluorescence, which indicated lateral diffusion of new membrane-bound fluorescent molecules, it was possible to get a quantitative measurement of membrane fluidity. This technique has been called "fluorescence recovery after photobleaching" (FRAP). These studies did not alter the membrane properties directly. Rather, they were designed to alter the fluorescent probe bound to specific membrane moieties. However, it is possible that, in some instances, the membrane was damaged by energy transfer from the fluorescent probe to the membrane components.

Indirect membrane effects also were demonstrated in a series of studies involving selective laser irradiation of the large mitochondria in cardiac myocytes. In these studies, exposure of a single mitochondrion either to a pulsed 532-nm frequency-doubled Nd:YAG laser or to the blue-green argon ion laser resulted in structural damage to the irradiated mitochondrion as well as an observed changed in cellular contractility. The cell often entered a state of uncoordinated contraction that could be shown to have a concomitant alteration in electrical activity of the outer cell membrane (Kitzes and Berns, 1979). The membrane effect was due either to a thermal or mechanical effect after absorption by the mitochondrion or to the release of intramitochondrial calcium that caused a change in membrane polarization.

One of the first direct laser scissors effects on the outer cell membrane was production of a transient membrane effect, thus allowing exogenous molecules to enter the cell. This was first described by Tsukakoshi *et al.* in 1984 as a method to allow selective DNA transfection of cells. This approach was subsequently adopted by Greulich and colleagues for a variety of studies on plant and animal cells (Weber *et al.*, 1988), and by Tao *et al.* (1987) for transfection of human cells in culture (Fig. 6). In these studies, the third harmonic 355-nm wavelength of a Nd:YAG laser was used to alter the cell membrane so that a gene correcting for hypoxanthine phosphoribosyltransferase (HPRT) deficiency could be incorporated into the genome of human fibrosarcoma cells. Transfection rates of 10^{-4} -10^{-3} were possible. The mechanism of this *optoporation* is not known, but because of the laser fluences used, it could have been due to multiphoton absorption. Subsequent studies were conducted on rice plant callus cells in culture (Guo *et al.*, 1995). In these studies, genes for kanamycin antibiotic resistance and beta glucuronidase were injected by laser into single cellus cells, from which an entire

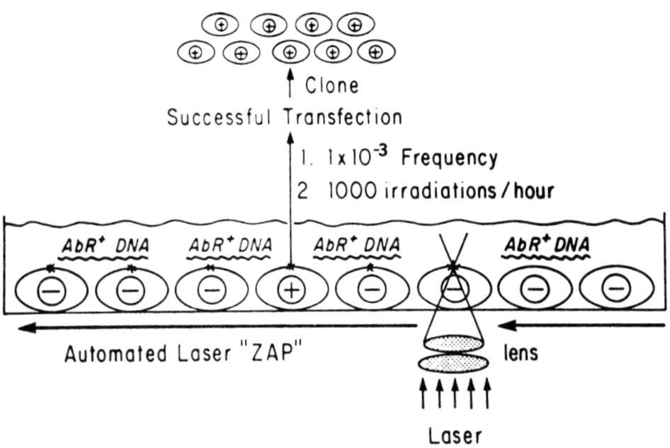

Fig. 6 Schematic representation of laser-mediated gene transfer. The cells (−) are irradiated with the focused laser beam in the presence of the plasmid DNA. The plasmid DNA, contained in the culture medium, is thought to be introduced into the cells through a very small hole momentarily made in the membrane by the laser. The transformants (+) are then selected and expanded to stable cell lines in selective medium.

rice plant was generated. All the cells in the regenerated plant contained the inserted genes.

The mechanism of optoporation is still not well understood. At the fluences, used, it is possible that the membrane alteration may have been an actual physical opening caused by a microplasma-generated shock wave, or it could have been a multiphoton-induced UV effect that transiently altered the membrane structure so that molecules entered the cell through a transient laser-induced optical pore.

In addition to the preceding studies in which the laser was used as a scissors on the cell membrane to facilitate the entry of molecules, the laser scissors has also been used to alter the cell membrane so that two cells in apposition could be fused together to form one cell (Wiegand-Steubing *et al.,* 1991). In this study a 366-nm UV beam from a nitrogen laser-pumped dye laser was used to alter a micron-size region of two myeloma cells that had been physically brought together by use of a 1.06-μm optical tweezers. (see Fig. 1). This study was one of the first examples showing the combined use of the laser tweezers and scissors.

D. Reproductive Biology

The introduction of *in vitro* fertilization (IVF) into clinical practice has changed the approach used for infertility. Two main areas have received special attention in the last few years: (*a*) the use of IVF for the treatment of male infertility, and (*b*) improvement of implantation to achieve a higher pregnancy rate. Several methods have been studied in these areas, and recent studies suggest that gamete manipulation may play a major role in solving both problems. Meticulous handling of gametes during such manipulations requires special tools. Laser microbeams offer potential advantages as accurate manipulating tools for cellular and subcellular organelles and, as such, were suggested and tested for gamete manipulations. In the 6 years since the introduction of the laser to the IVF laboratory, it has been tested for the following applications: (*a*) optical trapping to manipulate sperm and study new physiologic aspects of sperm motility, (*b*) laser zona drilling (LZD) to improve fertilization in the presence of abnormal sperm, and (*c*) laser assisted hatching (LAH) to improve implantation. Several commercial systems using various wavelengths dedicated to gamete manipulations have been developed and are undergoing clinical trials.

1. Sperm Manipulation

Laser-generated optical tweezers have been applied to manipulate sperm in two and three dimensions (Colon *et al.,* 1992; Schutze *et al.,* 1993; Tadir *et al.,* 1989, 1990). Initially, the CW Nd:YAG laser operating at 1064 nm was used to determine relative force generated by single spermatozoa that exhibited different swimming velocities and motility patterns (Tadir *et al.,* 1990). It was demonstrated that sperm with a zigzag pattern swim with more force than straight-swimming sperm. Other experiments conducted with a tunable CW Ti:sapphire laser

(700- to 800-nm wavelength) produced similar results (Araujo *et al.*, 1994; Westphal *et al.*, 1993).

Several studies have been performed to study the physiologic effects of laser trapping on sperm. Measurements of the relative sperm swimming force before and after the sperm encountered the cumulus cell mass of the egg determined that a significant increase in swimming force occurred after interaction with the cumulus mass (Westphal *et al.*, 1993). Relative force measurement of human sperm before and after cyropreservation demonstrated no significant difference when a yolk buffer freezing media was used as a cryoprotectant (Dantas *et al.*, 1995). In another study, the relative escape force of human epididymal sperm (aspirated microsurgically for IVF) was tested and compared to that of normal sperm with respect to the fertilizing potential of such sperm *in vitro*. Data suggested that the relative swimming force of the epidydimal sperm was significantly lower (60%) than that of ejaculated sperm, and that this was also reflected in the lower fertilization rate *in vitro* (Araujo *et al.*, 1994). In a recent study using the same system, it was determined that *in vitro* exposure of human sperm to pentoxifylline significantly increases swimming forces in normospermic and asthenospermic samples. This experiment confirmed that optical tweezers can provide an accurate determination of sperm force in experimental *in vitro* conditions. However, recent progress in other techniques of assisted fertilization (such as intracytoplasmatic sperm injection) may limit the use of the optical tweezers in reproductive medicine to studies of sperm physiology as opposed to actual sperm manipulation in clinical IVF (Patrizio *et al.*, 1996).

2. Oocyte Manipulation: Laser Zona Drilling (LZD)

Drilling holes in the zona pellucida (ZP) with a tunable dye laser at various wavelengths (266–532 nm) was first described in 1989 (Tadir *et al.*, 1989). Mouse, hamster, and primate oocytes were exposed to laser beams in a contact-free mode. The beam was delivered through the microscope objective, and the depth of incision was observed on a television monitor and controlled by a joystick-activated motorized stage (Fig. 7). This method is simple and accurate compared with conventional micromanipulations. The xenon chloride (XeCl) excimer laser (operating at 308 nm) in a similar nontouch configuration was applied to perform even more accurate incisions (Neev *et al.*, 1992). It can be applied for LZD and assisted hatching. The accuracy of this method enables the drilling of several neighboring apertures without causing visible damage to the delicate vitelline membrane of the egg. The krypton fluoride laser (operating at 248 nm) was applied to two cell mouse embryos to create a 2- to 4-μm opening in the zona pellucida (Blanchet *et al.*, 1992). It was concluded that by selecting the right parameters, the clean cuts did not intefere with blastocyst formation.

A different approach for drilling holes in the zona pellucida using glass pipettes or laser fibers in a contact mode, was suggested by several investigators. In these studies, the argon fluoride laser at 193 nm (Palanker *et al.*, 1991) Nd:YAG laser

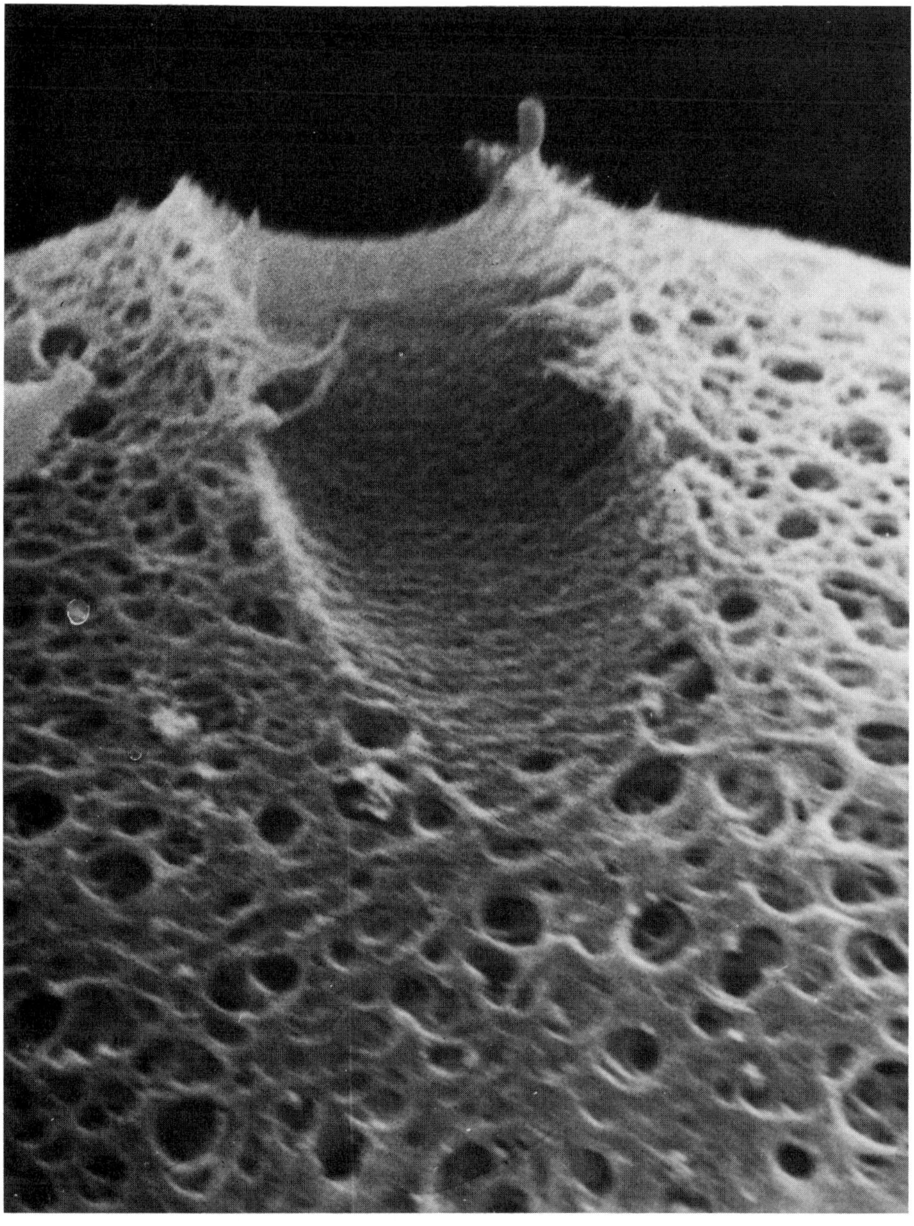

Fig. 7 Scanning electron micrograph of mouse oocyte with a narrow trench cut out of the zona pelucida by using a noncontact laser.

at 1060 nm (Coddington *et al.*, 1992), holmium (Ho):YAG laser at 2100 nm (Reshef *et al.*, 1993), and erbium (Er):YAG laser at 2940 nm (Feichtinger *et al.*, 1992) were applied to oocytes.

The excimer 193-nm laser was delivered to the mouse oocyte zona pellucida through a series of mirrors and a long focal length lens connected to an alumina silicate pipette. The glass pipette was pulled from capillaries with a 1-mm outer diameter to a tip of about 3- to 5-μm and filled with positive air pressure. Insemination at low sperm densities led to fertilization and further development to the blastocyst stage. The Ho:YAG laser was delivered through a 320-μm silica fiber tapered to 10- to 20-μm at its distal end to perform zona drilling in mouse oocytes before insemination *in vitro*. Of 222 laser-drilled oocytes 101 (45.5%) progressed to blastocyst stage or beyond after IVF, compared with 112 out of 246 (45%) in the nontreated controls. The authors concluded that although the laser-drilled holes did not improve fertilization, they did not impair the development to the blastocyst stage.

Initial success of a human pregnancy after Er:YAG-LZD created expectations for this system (Schiewe *et al.*, 1995). However, further studies using this system were shifted toward assisted hatching which is discussed in the next section.

In principle, the fiber delivery of laser for gamete manipulations is more cumbersome than the contact-free mode because conventional micromanipulating devices and disposable sterile equipment are needed. Contact delivery is required when the laser wavelength is shorter than ~200 nm or longer than ~2000 nm. Absorption of laser light by fluid is significant in this range, and an effective beam will not penetrate through the culture medium (Tadir *et al.*, 1993).

Laser-zona interaction and biological effects after contact-free laser exposure in the 200- to 2000-nm range have been studied by several groups.

A nitrogen-pumped dye laser operating at 440 nm and a pulse rate of 20 ppsec was focused through an epifluorescence port of an inverted microscope to produce 9- to 11-μm opening in the ZP of murine and bovine oocytes (Godke *et al.*, 1990). The success rate for accurate ZP cutting was 93% for murine and 100% for bovine oocytes.

In a previous study we demonstrated the influence of various physical parameters on the expected effects during gamete manipulations (Neev *et al.*, 1992). A series of laser microsurgery experiments was performed on nonviable discarded human oocytes that failed to fertilize in standard IVF treatment. Almost 400 of these failed unfertilized oocytes from 120 IVF cycles served as the experimental material. Oocytes were micromanipulated with two different excimer lasers (the 193-nm ArF and the 308-nm XeCl). Effects were video recorded and analyzed by computerized image processing and scanning electron microscopy (SEM). Ablation holes smaller than 1 μm were obtained in a reproducible fashion without causing any apparent damage to neighboring areas. This noncontact mode allowed for simultaneous viewing and cutting if proper laser parameters were chosen. Pulse energy and the beam focal plane position were shown to be the most critical parameters in determining the ablated spot diameters. It was concluded that excimer lasers of 308 nm operating in a short-pulse mode (15–250 nsec) are effective microsurgical tools for achieving removal, in a noncontact mode, of a portion of the zona pellucida. At this particular wavelength, the optical absorption is strong enough to allow selective interaction with the zona pellucida, yet weak enough to minimize accumulation of heat or explosive ablation. In addition, the 308-nm radiation can be delivered through slides, microscope objectives, optical fibers, and fluid or oil. It can facilitate the removal of accurate and highly reproducible material without the need for handling and maintaining a contact delivery system.

El-Danasouri *et al.* (1993b) used the 308-nm noncontact pulsed laser directed through a 100× quartz objective to perform LZD before insemination in mouse oocytes. The laser energy at the objective focal point was 0.4-μJ pulse with a spot diameter of 1 μm. The microscope stage was moved until the zona approached the laser tangentially, and a photoablation slit was made. Zonae-drilled oocytes were inseminated with low sperm concentrations (2×10^4 sperm/ml) and compared to two control groups consisting of zona intact oocytes inseminated with either similarly low sperm concentrations or normal sperm concentrations (2×10^6 sperm/ml). Laser manipulated oocytes showed more than a sixfold improvement in fertilization rate over that of nonmanipulated oocytes (65% vs 10.4%), and 94.9% of the zygotes developed to the two-cell stage. However, blastocyst formation in the laser-manipulated eggs was significantly lower than that of the control group inseminated with normal sperm concentration (68.5% vs 90.2% $p < 0.01$). The authors concluded that the 308-nm laser has potential as a simple noncontact drill for improving fertilization with low sperm count. However, further characterization of laser parameters is needed to improve embryo growth before application in human IVF. El Danassouri *et al.* (1993a) further investigated the possible effect of superoxide anion on the fertilization and cleavage rate in similar animal models. The idea was based on the principle that this compound is known to reduce intra- or extracellular free radicals, particularly singlet oxygen, that may be generated by laser irradiation. Results showed no effect on blastocyst formation.

Another study used a nitrogen laser (337 nm) delivered through an inverted microscope to provide a spot smaller than 1 μm in the noncontact mode (Ng *et al.*, 1993). The laser was operated at 2.5 μJ pulse with a repetition rate of 10 pulses/sec. A 10-μm opening was made in each ZP of mouse oocytes. The drilled oocytes were then exposed to microdroplets with murine sperm at 2×10^5 sperm/ml. Two sets of controls were used: LZD oocytes without insemination and IVF (similar sperm concentration) of cumulus-free oocytes. There was a significant improvement in fertilization and blastocyst formation at day 5 after LZD (89 of 158 [65.2%] compared with 46 of 127 [36.2%] $p < 0.001$), and implantation occurred after transfer of these embryos into surrogate females. The authors concluded that this laser is safe and effective for human IVF.

The possibility of DNA damage by UV radiation must be carefully considered in dealing with genetic material of gametes. It is not clear if significant UV absorption is taking place after propagation through barriers such as the ZP, basal membrane, and cytoplasm, especially with the low energy levels and the tangential superficial exposure that are used. In addition, certain factors such as the choice of solvent, the pH, the concentration of a solution, and even the temperature may alter the absorption characteristics of the medium and the target (Kochevar, 1989; Smith and Hanawalt, 1969).

Laufer *et al.* (1993) examined the safety and efficacy of the argon fluoride excimer laser (193 nm) by drilling the zona pellucida of mouse oocytes to improve fertilization capacity (Laufer *et al.*, 1993). Laser drilling significantly enhanced both fertilization and hatching rates over that of controls. Normal litters were obtained from the transfer of embryos developed from laser-drilled oocytes into pseudopregnant recipients. It appears unlikely that mutagenesis would be a problem at the low fluence used in zona manipulations, especially when used in a tangential approach (when few photons are scattered through the membrane).

3. Preembryo Manipulation: Laser-Assisted Hatching (LAH)

Micromanipulation of embryos before transfer into the uterus has been suggested to enhance implantation after IVF. This suggestion was based on observations in selected groups of patients that an artificial opening of the ZP (assisted hatching) enhanced implantation rate (Cohen, 1991) and accelerated the process of implantation as indicated by the early rise of hormonal markers such as luteal estradiol, progesterone, and human chorionic gonadotropin (HCG) (Liu *et al.*, 1993). The thickness and hardness of the ZP are probably some of the factors that play a role in this complex process. The accuracy and simplicity with which the laser can be utilized to open the ZP without causing visible damage to the ooplasm membrane have increased the interest of this approach.

Effects on embryonic development were evaluated after use of the 308-nm XeCl excimer laser (Neev *et al.*, 1993). Zonae of 8 to 16 cell mouse embryos were either lased-irradiated ($n = 189$), zona drilled with acidified Tyrode's solution ($n = 183$), or left zona intact ($n = 188$). Blastocyst formation (99–100%) was similar in the three groups. Hatching occurred earlier in the laser-treated embryos than in those of the control groups. These embryos actually hatched through the laser-ablated area. Significantly more embryos were hatching on days 4 and 7 in the conventionally drilled group than in the laser-treated group. However, implantation rates of morphologically normal laser-ablated embryos were not impaired when compared with the control embryos. Even though the 308-nm laser appears to be safe, the sensitivity of gamete manipulation using ultraviolet radiation has caused most investigators to focus their research on the IR region of the spectrum.

The Ho:YAG laser operating in the IR region (2100 nm) and delivered through a silicon fiber was applied on the ZP of 2- to 8-cell-stage mouse embryos to assist hatching (Reshef *et al.*, 1993). The rate of development to blastocyst stage or beyond and the rate of hatching were compared between the laser-treated and control embryos. Embryos were placed in phosphate-buffered saline under oil during laser exposure, and assessed 72 hr later. Of 49 (67%) laser-drilled embryos, 33 progressed to hatching compared to 36 of 82 (44%) in untreated controls ($p < 0.01$).

Schiewe *et al.* (1995) assessed the efficacy of the same wavelength generated from a holmium:yttrium scandian gallium garnet laser (Ho:YSGG) operating at 2100 nm in a pipette-free noncontact mode to assist hatching and sustain normal embryonic development. They tested the unit with a pulse duration of 250 μsec and pulse repetition rate at 10 Hz. Incisions in the zona were obtained using 10 mJ/pulse. Two-cell mouse embryos were recovered and assigned to LAH or control groups. Fewer ($P < 0.05$) embryos developed to the blastocyst stage in the control group (81%) the LAH compared with group (90%). The procedure was deemed simple and accurate.

Feichtinger *et al.* (1992) applied the Er:YAG, contact laser to mouse embryos, and subsequently to human embryos. Groups of 10 to 15 mouse embryos were placed under oil on two slides. A control slide was maintained on a warming stage while embryos on the other slide were subjected to the laser to produce holes in the ZP. Subsequently, embryos were assessed for the number developing to the blastocyst stage. There was no difference between the laser-treated mouse embryos and the untreated controls on days 1 and 2 of culture. On day 3, however, complete hatching was significantly enhanced in the laser-treated group [44 of 55 (80%)] compared with that of controls [17 of 58 (29.3%), $p = 0.0001$].

The same laser was used in a multicenter human study (Oburca *et al.*, 1994). Embryos obtained from 129 patients who previously experienced repeated implantation failures after IVF and embryo transfer were exposed to similar laser treatment for assisted hatching. During the procedure, embryos were held by negative pressure using a glass holding pipette, and ZP ablation was performed by depositing approximately 10 μJ in the contact mode. Five to eight pulses were employed to penetrate the ZP and create a 20- to 30-μm opening. An ongoing pregnancy rate of 36% (30 of 84 patients) and 29% (13 of 45 patients) was achieved in the two centers, which represented encouraging results considering the patient group studied. Preliminary results of an ongoing prospective randomized study in patients with an initial failed IVF attempt exhibited a 50% pregnancy rate (10 of 20 patients) in the LAH group in contrast to 44% (10 of 23 patients) in the group without LAH. Implantation rate per embryo in this preliminary study also was not significant [LAH (23.8%) vs control (21%)]. These results suggest that the laser is not detrimental to embryo survival.

An alternative IR diode laser operating in the contact-free mode at 1485 nm was introduced to the IVF laboratory in 1994 (Rink *et al.*, 1996). In several studies the beam was delivered through a 45× objective of an inverted microscope (2- to 4-μm spot diameter, 10–40 msec pulse, 0.5–1.2 mJ) to produce laser zona dissection in mouse zygotes (Germond *et al.*, 1995a; Rink *et al.*, 1996). One discharge was sufficient to drill an opening in the ZP ranging from 5- to 20-μm depending on laser power and exposure time. Of the drilled zygotes 70% developed to the blastocyst stage, which was comparable to that of the control group, and there was no evidence of thermal damage. The same colleagues further explored the effects of the same diode laser in another set of studies. Germond *et al.* (1995a) demonstrated that the energy needed to drill a hole of a given diameter is greater for mouse and human zygotes than for oocytes. Safety of the drilling procedure was demonstrated by the fact that there were 42 normal mice born following the procedure and 33 normal second-generation newborns produced by four males and four laser-treated females that were cross-mated. Various laser parameters were tested (irradiation time of 3–100 ms, and laser power 22–55 mW) to determine potential thermal damage. The authors concluded that the microdrilling procedure can generate standardized holes in mouse ZP without any visible side effects. Human studies using the same diode laser demonstrated an improved pregnancy rate after LAH of cryopreserved embryos (Germond *et al.*, 1995b).

IV. SUMMARY

In summary, we described the use of laser scissors and tweezers from three perspectives: (*a*) the historical background from which these two techniques evolved, (*b*) an understanding and lack of understanding of the mechanisms of interaction with the biological systems, and (*c*) the applications of the scissors and tweezers alone and in combination. As the technology improves and we gain a better understanding of how these two tools operate they will become even more useful in probing cell structure and function, as well as practically manipulating cells in genetics, oncology, and developmental biology.

References

Aist, J. R., and Berns, M. W. (1981). Mechanics of chromosome separation during mitosis in *Fusarium* (*Fungi imperfecti*): New evidence from ultrastructural and laser microbeam experiments. *J. Cell Biol.* **91,** 446–458.

Albrecht-Beuhler, G., and Bushnell, L. A. (1979). The orientation of centrioles in migrating 3T3 cells. *Exp. Cell Res.* **120,** 111–118.

Alexander, S. P., and Rieder, C. L. (1991). Chromosome motion during attachment to the vertebrate spindle: initial saltatory-like behavior of chromosomes and quantitative analysis of force production by nascent kinetochore fibers. *J. Cell Biol.* **113,** 805–815.

Araujo, E., Tadir, Y., Patrizio, P., Ord, T., Silber, S., Berns, M. W., and Asch, R. (1994). Relative force of human epididymal sperm correlated to the fertilizing capacity in vitro. *Fertil. Steril.* **62,** 585–590.

Ashkin, A. (1980). Applications of radiation pressure. *Science* **210,** 1081–1088.

Ashkin, A., and Dzeidzic, J. M. (1987). Optical trapping and manipulation of viruses and bacteria. *Science* **235,** 1517–1520.

Ashkin, A., and Dzeidzic, J. M. (1989). Internal cell manipulations using infrared laser traps. *Proc. Natl. Acad. Sci. USA* **86,** 7914–7918.

Ashkin, A., Dzeidzic, J. M., and Yamane, T. (1987). Optical trapping and manipulation of single cells using infrared laser beams. *Nature* **330,** 769–771.

Ashkin, A., Dzeidzic, J. M., Bjorkholm, J. E., and Chu, S. (1986). Observation of a single beam gradient force optical trap for dielectric particles. *Optic Lett.* **11,** 288–296.

Berns, M. W. (1968). Growth and morphogenesis during the post-embryonic development of the milliped. *Narceus annularis.* New York: Cornell University, Ithaca, Ph.D. Thesis.

Berns, M. W. (1974). Biological microirradiation. Englewood Cliffs, New Jersey: Prentice-Hall, pp. 152.

Berns, M. W. (1976). A possible two photon effect in vitro using a focused laser beam. *Biophys. J.* **16,** 973–977.

Berns, G. S., and Berns, M. W. (1982). Computer-based tracking of living cells. *Exp. Cell Res.* **142,** 103–109.

Berns, M. W., and Rounds, D. E. (1970). Cell surgery by laser. *Sci. Am.* **222,** 98–110.

Berns, M. W., and Salet, C. (1972). Laser microbeams for partial cell irradiation. *Int. Rev. Cytol.* **33,** 131–156.

Berns, M. W., Olson, R. S., and Rounds, D. E. (1969). *In vitro* production of chromosomal lesions using an argon laser microbeam. *Nature* **221,** 74–75.

Berns, M. W., Gamaleja, N., Olson, R., Duffy, C., and Rounds, D. E. (1970a). Argon laser microirradiation of mitochondria in rat myocardial cells in tissue culture. *J. Cell Physiol.* **76,** 207–214.

Berns, M. W., Ohnuki, Y., Rounds, D. E., and Olson, R. S. (1970b). Modification of nucleolar expression following laser microirradiation of chromosomes. *Exp. Cell Res.* **60,** 133–138.

Berns, M. W., Rattner, J. B., Brenner, S., and Meredith, S. (1977). The role of the centriolar region in animal cell mitosis: A laser microbeam study. *J. Cell Biol.* **72,** 351–367.

Berns, M. W., Chong, L. K., Hammer-Wilson, M., Miller, K., and Siemens, A. (1979). Genetic microsurgery by laser: Establishment of a clonal population of rat kangaroo cells (PTK_2) with a directed deficiency in a chromosomal nucleolar organizer. *Chromosoma* **73,** 1–8.

Berns, M. W., Aist, J., Edwards, J., Strahs, K., Girton, J., McNeill, P., Rattner, J. B., Kitzes, M., Hammer-Wilson, M., Liaw, L.-H., Siemens, A., Koonce, M., Peterson, S., Brenner, S., Burt, J., Walter, R., van Dyk, D., Coulombe, J., Cahill, T., and Berns, G. S. (1981). Laser microsurgery in cell and developmental biology. *Science* **213,** 505–513.

Berns, M. W., Wright, W. H., Tromberg, B. J., Profeta, G. A., Andrews, J. J., and Walter, R. J. (1989). Use of a laser-induced optical force trap to study chromosome movement on the mitotic spindle. *Proc. Natl. Acad. Sci. USA* **86,** 4539–4543.

Berns, M. W., Wright, W. H., and Wiegand-Steubing, R. (1991). Laser microbeam as a tool in cell biology. *Int. Rev. Cytol.* **129,** 1–44.

Berns, M. W., Liang, H., Sonek, G., and Liu, Y. (1994). Micromanipulation of chromosomes using laser microsurgery (optical scissors) and laser-induced optical forces (optical tweezers). Cell Biology: A Laboratory Handbook. New York: Academic Press, pp. 217–227.

Berns, M. W., Hamkalo, B. H., and Beissman, H., unpublished.

Bessis, M., Gires, F., Mayer, G., and Nomarski, G. (1962). Irradiation des organites cellulaires a l'aide d'um laser rubis. *C. R. Acad. Sci.* **225,** 1010–1012.

Blanchet, B. B., Russell, J. B., Fincher, C. R., and Portman, M. (1992). Laser micromanipulation in the mouse embryo: A novel approach to zona drilling. *Fertil. Steril.* **57,** 1337–1347.

Calmettes, P. P., and Berns, M. W. (1983). Laser-induced multiphoton processes in living cells. *Proc. Natl. Acad. Sci. USA* **80,** 7197–7199.

Coddington, C. C., Veeck, L. L., Swanson, R. J., Kaufman, R. A., Lin, J., Simonetti, S., and Bocca, S. (1992). The YAG laser used in micromanipulation to transect the zona pellucida of hamster oocytes. *J. Assist. Reprod. Genet.* **9,** 557–563.

Cohen, J. (1991). Assisted hatching of human embryos. J. IVF and ET 8:179–190.

Colon, J. M., Sarosi, P., McGovern, P. G., Ashkin, A., Dziedzic, J. M., Skurnick, J., Weiss, G., and Bonder, E. M. (1992). Controlled micromanipulation of human spermatozoa in three dimensions with an infrared laser optical trap: Effect on sperm velocity. *Fertil. Steril.* **57,** 695–698.

Daniel, J. C., and Takahashi, K. (1965). Selective laser destruction of rabbit blastomeres and continued cleavage of survivors *in vitro. Exp. Cell Res.* **39,** 475–479.

Dantas, Z., Araujo, E., Berns, M. W., Tadir, Y., Schell, M. W., and Stone, S. C. (1995). Effect of freezing on the relative escape force of sperm as measured by laser optical trap. *Fertil. Steril.* **63,** 185–188.

Denk, W. (1996). Two-photon excitation in functional biological imaging. *J. Biomed. Optics* **1,** 296–304.

Denk, W., Strickler, J. H., and Webb, W. W. (1990). Two-photon laser scanning microscopy. *Science* **248,** 73–76.

Denk, W., Delaney, K. P., Gelpenn, A., Kleinfeld, D., Strowbridge, B. W., Tank, D. W., and Yuste, Y. (1994). Anatomical and functional imaging of neurons using 2-photon laser scanning microscopy. *J. Neurosci. Methods* **54,** 151–162.

Djabali, M., Nguyen, C., Bianno, I., Oostra, B., Mattei, M., Ikeda, J., and Jordan, B. (1991). Laser microdissection of the fragile x region: identification of cosmid clones and of conserved sequences in this region. *Genomics* **10,** 1053–1060.

El Danasouri, I., Gebhardt, J., and Westphal, L. M. (1993a). Superoxide dismutase improves the fertilization rate of mouse oocytes micromanipulated with 308-nm excimer laser and fertilized with low sperm concentration. *J. Assist. Repord. Prog. Suppl.* Abstract No. 199.

El Danasouri, I., Westphal, L. M., Neev, Y., Gebhardt, J., Louie, D., Tadir, Y., and Berns, M. W. (1993b). Assisted fertilization of mouse oocytes using a 308-nm excimer laser microbeam to open the zona pellucida. *J. Assist. Reprod. Prog. Suppl.* Abstract No. 228.

Feichtinger, W., Strohmer, H., Fuhrberg, P., Radivojevic, K., Antoniori, S., Pepe, G., and Versaci, C. (1992). Photoablation of oocyte zona pellucida by erbium:YAG laser for *in vitro* fertilization in severe male infertility. *Lancet* **339,** 811.

Germond, M., Nocera, D., Senn, A., Rink, K., Delacretaz, G., and Fakan, S. (1995a). Microdissection of mouse and human zona pellucida using a 1.48-μm diode laser beam: Efficacy and safety of the procedure. *Fertil. Steril.* **64,** 604–611.

Germond, M., Senn, A., Rink, K., Delacretaz, G., and De Grandi, P. (1995b). Is assisted hatching of frozen-thawed embryos enhancing pregnancy outcome in patients who have several previous nidation failures? *J. fur Fertilitat und Reproduktion* **3,** 41.

Godke, R. A., Beetem, D. D., and Burleigh, D. W. (1990). A method for zona pellucida drilling using a compact nitrogen laser. Presented at the VII World Congress on Human Reproduction, June 26–July 1, Abstract No. 258.

Greulich, K. O., and Leitz, G. (1994). Light as microsensor and micromanipulator: Laser microbeams and optical tweezers. *Exp. Tech. Phys.* **40,** 1–14.

Greulich, K. O., and Wolfrum, J. (1989). Ber. Bunsenges. *Phys. Chem.* **93,** 245.

Guo, Y., Liang, H., and Berns, M. W. (1995). Laser-mediated gene transfer in rice. *Physiologia Plantarum* **93,** 19–24.

He, W. (1995). Laser microdissection and its application to be human tuberous sclerosis 1 gene region on chromosome 9q 34. 1995. Irvine, California: University of California, Ph.D. Thesis.

He, W., Liu, Y., Smith, M., and Berns, M. (1997). Laser microdissection for generation of a human chromosome region-specific library. *Microsc. Microanal.* **3,** 47–52.

Kitzes, M. C., and Berns, M. W. (1979). Electrical activity of rat myocardial cells in culture: La^{+++}-induced alterations. *Am. J. Physiol.: Cell Physiol.* **6,** C87–C95.

Kochevar, I. E. (1989). Cytotoxicity and mutagenicity of excimer laser radiation. *Lasers Surg. Med.* **9,** 440–445.

Konig, K., Liang, H., Berns, M. W., and Tromberg, B. J. (1995). Cell damage by near-IR microbeams. *Nature* **377,** 20–21.

Konig, K., Liang, H., Berns, M. W., and Tromberg, B. J. (1996). Cell damage in near-infrared multimode optical traps as a result of multiphoton absorption. *Optics Lett.* **21,** 1090–1092.

Koonce, M. P., Cloney, R. A., and Berns, M. W. (1984). Laser irradiation of centrosomes in newt eosinophils: Evidence of centriole role in motility. *J. Cell Biol.* **98,** 1999–2010.

Koppel, D. E., Axelrod, D., Schlessinger, J., Elson, E. L., and Web, W. W. (1976). Dynamics of fluorescence marker concentration as a probe of mobility. *Biophys. J.* **16,** 1315–1329.

Laufer, N., Palanker, D., Shufaro, Y., Safran, A., Simon, A., and Lewis, A. (1993). The efficacy and safety of zona pellucida drilling by a 193-nm excimer laser. *Fertil. Steril.* **59,** 889–895.

Liang, H., Vu, K. T., Krishnan, P., Trang, T. C., Shin, D., Kimel, S., and Berns, M. W. (1996). Wavelength dependence of cell cloning efficiency after optical trapping. *Biophys. J.* **70,** 1529–1533.

Liang, H., Wright, W. H., Cheng, S., He, W., and Berns, M. W. (1993). Micromanipulation of chromosomes in PTK2 cells using laser microsurgery (optical scalpel) in combination with laser-induced optical force (optical tweezers). *Exp. Cell Res.* **204,** 110–120.

Liang, H., Wright, W. H., Rieder, C. L., Salmon, E. D., Profeta, G., Andrews, J., Liu, Y., Sonek, G. J., and Berns, M. W. (1994). Directed movement of chromosome arms and fragments in mitotic newt lung cells using optical scissors and optical tweezers. *Exp. Cell Res.* **213,** 308–312.

Liu, H. C., Noyse, N., Cohen, J., Rosenwaks, Z., and Alikani, M. (1993). Assisted hatching facilitates earlier implantation. *Fertil. Steri.* **60,** 871–875.

Liu, Y., Cheng, D. K., Sonek, G. J., Berns, M. W., and Tromberg, B. J. (1994). Microfluorometric technique for the determination of localized heating in organic particles. *Appl. Physics Lett.* **65,** 919–921.

Liu, Y., Sonek, G. J., Berns, M. W., Konig, K., and Tromberg, B. J., (1995). Two-photon excitation in continuous-wave infrared optical tweezers. *Optics Lett.* **20,** 2246–2248.

McNeill, P. A., and Berns, M. W. (1981). Chromosome behavior after laser microirradiation of a single kinetochore in mitotic PTK_2 cells. *J. Cell Biol.* **88,** 543–553.

Monajembashi, S., Cremer, C., Cremer, T., Wolfrum, J., and Greulich, K. O. (1986). Microdissection of human chromosomes by laser microbeam. *Exp. Cell Res.* **167,** 262–265.

Moreno, G., Lutz, M., and Bessis, M. (1969). Partial cell irradiation by ultraviolet and visible light: Conventional and laser sources. *Int. Rev. Exp. Path.* **7,** 99–137.

Neev, J. Gonzales, A., Licciardi, F., Alikani, M, Tadir, Y., Berns, M. W., and Cohen, J. (1993). A contact-free microscope delivered laser ablation system for assisted hatching of the mouse embryo without the use of a micromanipulator. *Human Reprod.* **8,** 939–944.

Neev, J., Tadir, Y., Ho, P., Asch, R. H., Ord, T., and Berns, M. W. (1992). Microscope-delivered UV laser zona dissection: Principles and practices. *J. Assist. Reprod. Genet.* **9,** 513–523.

Ng, S. C., Liow, S. L., Schutze, K., Vasuthevan, S., Bongso, A., and Ratnam, S. S. (1993). The use of ultraviolet microbeam laser zona dissection in the mouse. *J. Assist. Reprod. Prog. Suppl.* Abstract No. 273.

Oburca, A., Strohmer, H., Sakkas, D., Menezo, Y., Kogosowski, A., Barak, Y., and Feichtinger, W. (1994). Use of laser in assisted fertilization and hatching. *Human Reprod.* **9,** 1723–1726.

Oh, D. H., Stanley, R. J., Lin, M., Hoeffler, W., Boxer, S., Berns, M., and Bauer, E. (1997). Two-photon excitation of 4′-hydromethyl-4,5,8-trimethylpsoralen. *Photochem. Photobiol.* **65,** 91–95.

Palanker, D., Ohad, S., Lewis, A., Simon, A., Shenkar, J., Penchas, S., and Laufer, N. (1991). Technique for cellular microsurgery using the 193-nm excimer laser. *Lasers Surg. Med.* **11,** 580–586.

Patrizio, P., Liu, Y., Sonek, G., Berns, M. W., and Tadir, Y. (1996). Effect of pentoxyfyllin on the intrinsic force of human sperm. Presented at the American Academy of Andrology, Minneapolis, April 25–29.

Peterson, S. P., and Berns, M. W. (1980). The centriolar complex. *Int. Rev. Cytol.* **64,** 81–106.

Rattner, J. B., and Berns, M. W. (1974). Light and electron microscopy of laser microirradiated chromosomes. *J. Cell Biol.* **62,** 526–533.

Reshef, E., Haaksma, C. J., Bettinger, T. L., Haas, G. G., Schafer, S. A., and Zavy, M. T. (1993). Gamete and embryo micromanipulation using the holmium:YAG laser. *Fertil. Steril. Program Suppl.* P-016, S88.

Rink, K., Delacretaz, G., Salathe, R. P., Senn, A., Nocera, D., Germond, M., Fakan, S. (1994). Diode laser microdissection of the zona pellucida of mouse oocytes. *Biomed. Optics* **2134A,** 53.

Rink, K., Delacretaz, G., Salathe, R. P., Senn, A., Nocera, D., Germond, M., De Garnadi, P., and Fakan, S. (1996). Noncontact microdrilling of mouse zona pellucida with an objective delivered 1.48-μm diode laser. *Lasers Surg. Med.* **18,** 52–62.

Saks, N. M., Zuzdo, R., and Kopac, M. J. (1965). Microsurgery of living cells by ruby laser irradiation. *Ann. N. Y. Acad. Sci.* **122,** 695–712.

Salet, C., Moreno, G., and Vinzens, F. (1979). A study of beating frequency of a single myocardial cell. *Exp. Cell Res.* **120,** 25–29.

Schiewe, M. C., Neev, Y., Hazeleger, N. L., Balmaceda, J. P., Berns, M. W., and Tadir, Y. (1995). Developmental competence of mouse embryos following zona drilling using a noncontact Ho:YSSG laser system. *Human Reprod.* **10,** 1821–1824.

Schutze, K., Clemeny-Sengewald, A., and Berg, F. D. (1993). Laser zona drilling and sperm transfer into the perivitelline space. *Human Reprod.* **8,** 390.

Smith, K. C., and Hanawalt, P. C. (1969). Molecular Photobiology. Academic Press, New York, 230 pp.

Tadir, Y., Wright, W. H., Vafa, O., Ord, T., Asch, R., and Berns, M. W. (1989). Micromanipulation of sperm by a laser-generated optical trap. *Fertil. Steril.* **52,** 870–873.

Tadir, Y., Wright, W. H., Vafa, O., Ord, T., Asch, R. H., and Berns, M. W. (1990). Force generated by human sperm correlated to velocity and determined using a laser trap. *Fertil. Steril.* **53,** 944–947.

Tadir, Y., Neev, J., Ho, P., and Berns, M. W. (1993). Lasers for gamete micromanipulation: Basic concepts. *J. Assist. Reprod. Genet.* **10,** 121–125.

Tao, W., Wilkinson, J., Stanbridge, E. J., and Berns, M. W. (1987). Direct gene transfer into human cultured cells facilitated by laser micropuncture of the cell membrane. *Proc. Natl. Acad. Sci. USA* **84,** 4180–4184.

Tchakotine, S. (1912). Die mikrikopische Strahlenstrich methode, eine Zelloperationsmethode. *Biol. Zentralbl.* **32,** 623.

Tsukakoshi, M., Kurata, S., Nomiya, Y., Ikawa, Y., and Kasuya, T. (1984). A novel method of DNA transfection by laser microbeam surgery. *Appl. Phys. B* **35,** 135–140.

Vorobjev, I. A., Liang, H., Wright, W. H., and Berns, M. W. (1993). Optical trapping for chromosome manipulation: A wavelength dependence of induced chromosome bridges. *Biophys. J.* **64,** 533–538.

Weber, G., and Greulich, K. O. (1992). Manipulation of cells, organelles, and genomes by laser microbeam and optical trap. *Int. Rev. Cytol.* 133:1.

Weber, G., Monajembashi, S., Greulich, K. O., and Wolfrum, J. (1988). Injection of DNA into plant cells with a UV laser microbeam. *Naturwissenshaften* **75,** 35.

Westphal, L., El-Danasouri, I. E., Shimizu, S., Tadir, Y., and Berns, M. W. (1993). Exposure of human sperm to the cumulus oophorus results in increased relative force as measured by a 760-nm laser optical tram. *Human Reprod.* **8,** 1083–1086.

Wiegand-Steubing, R., Cheng, S., Wright, W. H., Numajiri, Y., and Berns, M. W. (1991). Laser-induced cell fusion in combination with optical tweezers: The laser cell fusion trap. *Cytometry* **12,** 505–510.

Zirkle, R. E. (1957). Partial cell irradiation. *Adv. Biol. Med. Phys.* **5,** 103–146.

Zirkle, R. E. (1970). Ultraviolet microbeam irradiation of newt cell cytoplasm. *Rad. Res.* **41,** 516–537.

CHAPTER 6

Optical Force Microscopy

Andrea L. Stout[1] and Watt W. Webb
School of Applied and Engineering Physics
Cornell University
Ithaca, New York 14853

I. Introduction

One goal of cell biology research is to observe individual cell components in their native environment. Nondamaging, high-resolution techniques for imaging molecular structures in aqueous buffer continue to be developed and improved. Scanning probe microscopy is emerging as an important technique for macromolecular imaging and manipulation, enabling investigators not only to image structures but also to examine their mechanical properties. In particular, use of the atomic force microscope (AFM) has become more and more prevalent for imaging biological surfaces.

Applications of the AFM now range from acquiring topographic maps of macromolecular assemblies and cell surfaces to exploring the forces involved in

[1] Present address: Department of Physics and Astronomy, Swarthmore College, Swarthmore, Pennsylvania 19081-1397

Copyright © 1998 by Academic Press. All rights of reproduction in any form reserved.
0091-679X/98 $25.00

maintaining intermolecular interactions (for a review see Lal and John, 1994). Although a success in many cases, the AFM is not always a convenient tool for exploring biological systems. One promising alternative, which is more compatible with optical microscopy than the AFM, is a scanning probe microscope built around the single-beam gradient optical trap, or optical tweezers. The optical tweezers have become established not only as a tool for manipulating microscopic particles, but as a sensitive force transducer as well. Because the tweezers act on small dielectric particles in a manner similar to that of a linear spring, they can serve as the foundation of a scanning probe microscope we refer to as the optical force microscope (OFM). The OFM provides an alternate means of exploring surfaces and intermolecular interactions, applications typically reserved for the AFM.

We here consider two broad categories of biological applications of scanning probe microscopes. The first is topographic imaging of surfaces and objects. In general, this type of application requires that a small probe be scanned in a raster pattern across a sample (with or without feedback) while deflections of the probe are monitored with high precision. In principle, the resolution limit in this type of imaging is determined by the size and shape of the probe as well as the probe's thermal kinetic energy. Included in this category are those experiments aimed toward mapping interaction forces rather than simply topographic information (Berger *et al.,* 1995; Frisbie *et al.,* 1994).

The second category encompasses experiments designed to investigate the interaction forces between individual molecules. Recent observations of nanonewton rupture forces for single pairs of molecules have utilized the small probe size and force sensitivity of the AFM. In these experiments "ligands" are attached to the AFM tip while "receptors" are attached to solid support. The tip is then brought into contact with an immobilized receptor, thereby presenting a few (or just one) molecules to it. The strength of the interaction can be measured by determining the force required to rupture adhesion between probe and surface. Examples of interactions studied in this manner are the streptavidin–biotin interaction (Florin *et al.,* 1994), an antibody–antigen interaction (Hinterdorfer *et al.,* 1996), and the interaction between DNA base pairs (Boland and Ratner, 1995; Lee *et al.,* 1994). The OFM extends the potential range of such measurements by several orders of magnitude into the single piconewton (pN) range.

II. AFM-like Applications

A. Optical Force Microscope

The use of optical tweezers as the basis for a scanning probe microscope was developed by Ghislain and Webb (1993). In a system very similar to that of an AFM, a calibrated optical trap replaces the AFM cantilever, and a particle in the trap serves as the tip. In this arrangement, an optically transparent sample is scanned beneath a probe particle held in close proximity by the optical tweezers.

Topographic features on the surface of the sample displace the probe relative to the center of the optical trap. These movements of the probe result in modulations of the intensity pattern of forward-scattered laser light, which is continuously monitored by a photodiode. These intensity modulations are analogous to those produced on a split photodiode by a laser beam reflecting off the back of an AFM cantilever. They are sent to a computer for display and analysis, which yields a two-dimensional image reflecting the topography of the sample. Like an AFM, the OFM is capable of operating with or without feedback to maintain a constant distance between the probe and the sample. Ghislain demonstrated the feasibility of such a system by imaging a series of AFM calibration grids cast in a transparent polymer as well as a network formed by drying small polystyrene beads on a substrate (Fig. 1 and 2). With this system, features as small as 20 nm could be imaged (Ghislain and Webb, 1993). Although thermal fluctuations of the probe particle prevent the OFM's spatial resolution from approaching that of an AFM, such an apparatus could be especially useful in combination with optical microscopy and for specimens that cannot tolerate the larger probe forces of the AFM.

B. Intermolecular Force Measurements

In the OFM the application of optical tweezers to the measurement of forces between molecules can parallel that of the AFM. In such experiments, the probe is coated with a few copies of a molecule while another molecule (with which the first molecule interacts) is linked at low densities to a surface (usually a glass slide or coverslip) whose position can be precisely controlled. In such applications, the ability of the optical tweezers to function as a highly sensitive force transducer is exploited. Although the AFM utilizes cantilevers with stiffnesses ranging from 10^2 to 10^{-3} Newtons/meter (N/m), the stiffness of the optical trap is generally 1 to 2 orders of magnitude lower than that of the softest AFM cantilever, enabling the measurement of forces as small as 0.5 pN (again, thermal fluctuations in the probe position, not instrument noise, place a lower limit on measurable forces). There is also an upper limit on the force measurable with the optical trap, known as the escape force. An external force greater than the escape force applied to a trapped particle will cause it to be ejected from the trap. A good optical trap will have an escape force on the order of 50 pN for a 1-μm probe particle. These factors suggest that the AFM is better suited than the OFM for exploring strong interactions such as hydrogen bonds (Israelachvili, 1992) or especially strong noncovalent bonds, whereas the optical tweezers are more suited for the measurement of forces associated with weaker, reversible interactions. For example, the force required to rupture a single antibody–antigen bond has been shown to approximate 240 pN (Hinterdorfer *et al.*, 1996); that needed to break the streptavidin-biotin bond approximates 180 pN. Such forces are, in general, inaccessible to the optical tweezers. In contrast, the forces exerted by individual motor molecules such as kinesin and myosin have been measured by optical

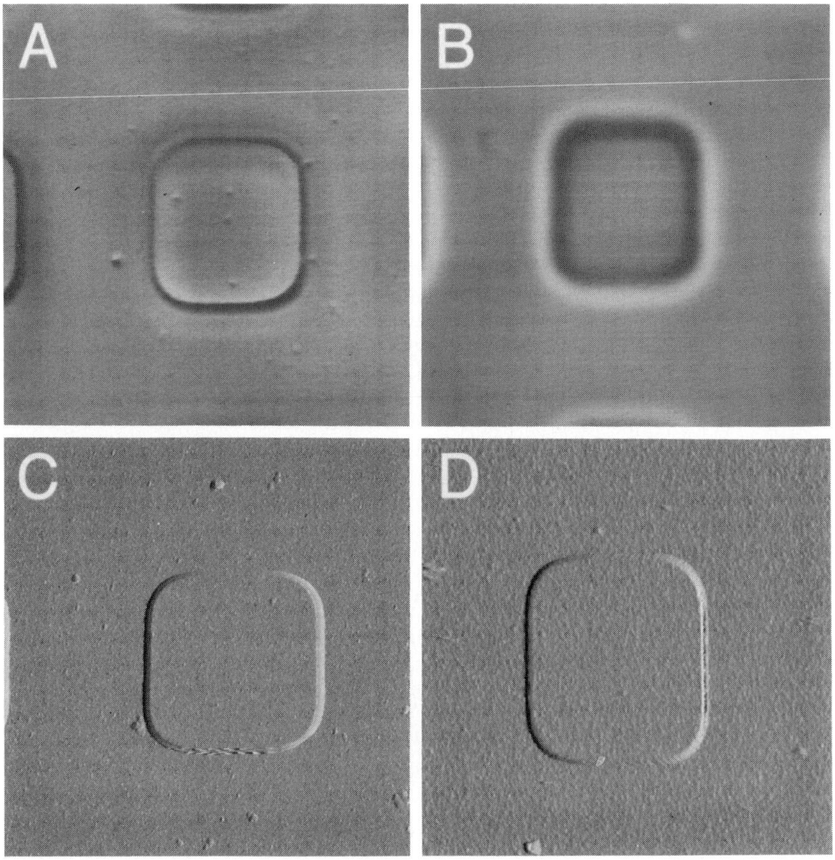

Fig. 1 A comparison of images of a calibration grid reproduced in Cibatool™ polymer (5 μm on a side, 180 nm high). (A) Image obtained with the OFM. The sample was immersed under water containing 0.03% casein and imaged with a 1.0-μm diameter polystyrene sphere. Scan rate was 2.0 Hz, and the total area is 13.5 × 13.5 μm. (B) Image collected as in (A) but with no probe particle. For comparison, (C) and (D) depict AFM images of the same grid. (C) Image obtained with a Nanoscope III Multimode AFM (Digital Instruments, Inc.) in air using "tapping mode." (D) AFM image obtained using contact mode with sample under water with 0.03% casein. Reproduced with permission from Ghislain (1994).

tweezers to range from around 7 to 9 pN, which is substantially smaller (Finer *et al.,* 1994; Svoboda *et al.,* 1993).

Although the optical tweezers cannot directly apply a force greater than the escape force, it is possible to channel the available force in a way that permits the exploration of interactions in the 50–200 pN range, which is normally accessible only to the AFM. By selecting a geometry that allows the probe particle to act

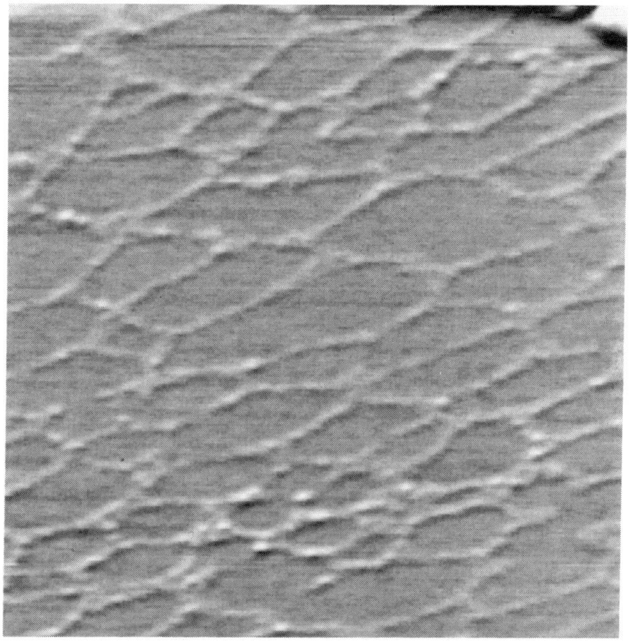

Fig. 2 OFM image of 250-nm polystyrene beads fixed to a coverslip in a meshlike aggregation pattern by drying from a methanol suspension. The scan area is approximately 14 × 17 μm. Reproduced with permission from Ghislain (1994).

as a lever arm, the force applied by an optical tweezers can be amplified by a factor as high as 8. This can be accomplished by using the configuration shown in Fig. 3. Here, the ligand-coated probe is scanned over a receptor-coated surface. The probe-to-surface distance must be kept small enough to allow frequent sampling of the surface by the probe via Brownian motion. When a bond forms, creating a tether between probe and surface, the probe will pivot about this tether as the scan continues, eventually being pressed against the surface. As soon as this firm contact with the surface occurs, the tangentially applied trap force is amplified by a factor of $1/\sin\theta$ as shown, due to the presence of a normal force between probe and surface: the shorter the tether, the greater the force amplification. Examples of such data obtained by using beads coated with several immunoglobulin G (IgG) molecules and coverslips sparsely coated with protein A from *Staphylococcus aureus* are shown in Fig. 4. Although this geometry offers a substantial mechanical advantage, it is highly sensitive to the presence of nonspecific interactions due to the nonnegligible contact area between probe and surface. It is therefore necessary to take measures to prevent nonspecific interactions as much as possible; this topic will be treated in more detail later in this chapter.

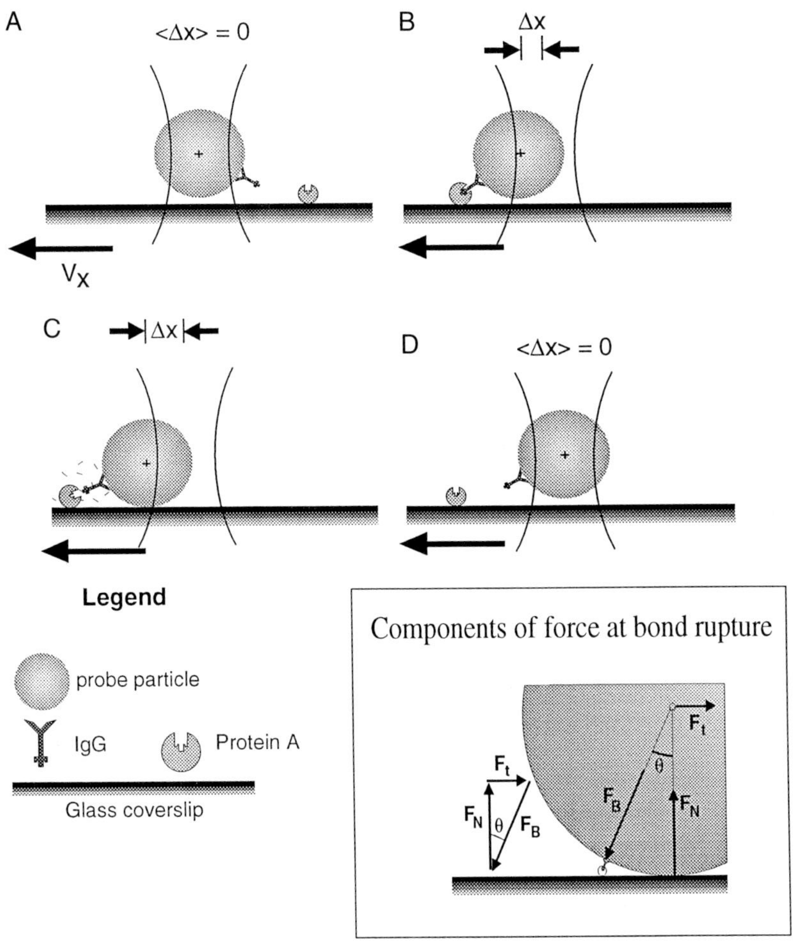

Fig. 3 Cartoon illustrating the use of an optical trap for intermolecular force measurements. (A) The probe with attached ligand is brought near a coverslip with attached receptors. The coverslip is scanned in the x direction at velocity v_x while the trap is held stationary. Displacement of the bead from the trap center, Δx, is monitored with a quadrant photodiode. Before binding, $\Delta x = 0$. (B) Upon binding of ligand to receptor, the bead begins to move with the coverslip at velocity nearly equal to v_x and a force $F(\Delta x)$ is exerted on the bead. (C) At the point of rupture, the force exerted by the trap via the bead is sufficient to overcome the intermolecular bond, and the link between bead and moving surface is destroyed. (D) The bead returns to its equilibrium position at the center of the trap and no longer moves with the surface. Inset: The force exerted by the optical trap at the point of breakage, $F(\Delta x)$, is amplified by a factor $1/\sin\,\theta$ as shown, due to the presence of a normal force between bead and surface.

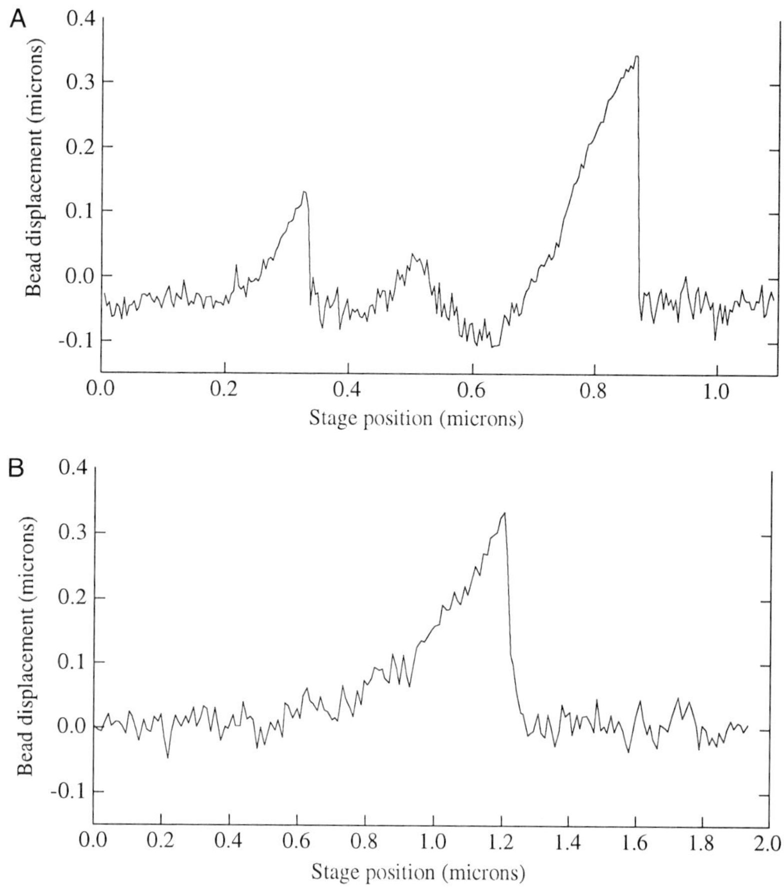

Fig. 4 Examples of signals obtained with the experimental protocol illustrated in Fig. 3. (A) Data obtained using 1.0-μm diameter polystyrene beads with an average of 10 bovine IgG molecules covalently attached. A trapped bead was scanned along a coverslip with a sparse population of covalently lined protein A. (B) Data obtained as in (A) but using 2.6-μm beads.

III. Experimental Design

Any scanning probe microscope requires five essential elements: (*a*) a probe that can be easily manipulated and that has desirable shape and interaction properties; (*b*) a means of scanning the sample or the probe; (*c*) a sensitive detector for determining the position of the probe relative to some equilibrium position; (*d*) proper calibration of the system so that the raw signal (time-dependent voltage or voltages) can be converted into the parameter of interest (position or force); and (*e*) signal processing equipment that will acquire signals with sufficient speed and allow for their display and analysis. These design elements are described in the following discussion; details on setting up a microscope with optical tweezers are given in previous chapters.

A. Probe Selection

In both types of experiment (imaging and force measurement), probe selection must be based on a compromise between several factors: maximizing trapping efficiency and resolution while minimizing surface interactions and thermal noise. Up to a point, the gradient force necessary for a stable single-beam optical trap increases with the size of the probe particle. Because of this, the root-mean-square thermal noise due to Brownian motion of the probe increases as particle size decreases; thus, larger probes result in a larger signal-to-noise ratio than do small probes. However, the desirable qualities of large probes are in direct conflict with the improvement in spatial resolution and minimization of nonspecific interactions that result from smaller probe particles. In addition, the optical characteristics of the material comprising the probe particle must be considered. The magnitude of the gradient force is also determined by the ratio of the probe's refractive index to that of the suspending medium (m): $m = n_{probe}/n_{medium}$. For particles that are small compared to the wavelength of trapping light (those categorized as Rayleigh scatterers), it can be shown that the gradient force is proportional to the particle radius a and the polarizability α, where

$$\alpha = n_{medium}^2 \frac{m^2 - 1}{m^2 + 2} a^3$$

(Visscher and Brakenhoff, 1992a). For larger particles, however (those in the Mie size regime, $a \gg \lambda$), trap strength increases with m only until an optimal value of $m \approx 1.25$ is reached (Visscher and Brakenhoff, 1992b). Increasing m beyond this value leads to diminished trapping strength due to an increase in the Mie regime equivalent of the scattering force. Although optical properties of the probe particle are important, they are not the only consideration when choosing a material. Because colloidal particles vary in their adhesive properties, it may be necessary to choose a probe with a suboptimal refractive index in order to minimize nonspecific interactions of the probe with the surface.

For optical tweezers formed by overfilling the back aperture of a high numerical aperture (NA) objective (1.3–1.4), theoretical and experimental investigations have shown that axial and radial trapping efficiencies tend to increase with the size of the particle (Wohland *et al.*, 1996; Wright *et al.*, 1994). For the commonly used 1064-nm light from a continuous-wave neodynium:yttrium aluminum garnet (Nd:YAG) laser, such conditions lead to a beam waist diameter of ~0.8 μm. Several investigations of trapping efficiency as a function of particle size and index of refraction have been conducted (Felgner *et al.*, 1995; Wright *et al.*, 1994) with spherical beads. In general, polystyrene (latex) spheres (n = 1.57), are more strongly trapped than amorphous silica (n = 1.45) spheres. This would appear to imply that larger polystyrene beads are the probe of choice, but one must also consider that polystyrene beads tend to exhibit stronger nonspecific interactions with surfaces (especially if they are uncharged) and have also been shown to have "hairy" surfaces with long tendrils of polymer, which often can tether the beads tightly to a substrate (Dabros *et al.,* 1994).

One way of improving spatial resolution and minimizing surface interactions without compromising particle size is to use elongated probe particles whose volume is comparable to that of the 1-μm spheres but whose contact area is much smaller. Such probes were used by Ghislain and Webb (1993) for imaging a polymer-coated surface. In preliminary experiments, glass shards approximately 3-μm long and 1-μm wide, with pointed tips, were held in an optical trap and scanned over surfaces. To be practical, however, elongated probes will have to be carefully designed and fabricated with an eye toward reproducible shape and the ability to be trapped in a stable manner (elongated particles tend to spin when held in an optical trap). As a means of trapping elongated probes, Ghislain (1994) proposed a pair of superimposed trapping beams whose foci are slightly offset in the axial direction.

Even a probe with a small radius of curvature will display adhesive behavior when brought into close contact with a surface. In pure water these adhesive forces are relatively weak, but at the salt concentrations (0.1 *M*) used for most experiments with biological molecules, they can present a significant obstacle. Such a concentration of ions generally allows very strong, permanent adhesion of colloidal particles to surfaces. Under these conditions, screening of repulsive electrical double-layer forces allows the strong, short-range dynamic Van der Waals attraction to dominate for distances less than 2 nm (Israelachvili, 1983). Such interference usually can be overcome by including in the medium an agent that can prevent this strong attraction, either via steric hinderance or an alteration of the surface potentials. The most effective blocking agent is highly dependent on the system in question; however, several investigators have reported success with the milk protein, casein (Stout and Webb, 1997; Svoboda *et al.,* 1993). Casein consists of four species, α_{s1}, α_{s2}, β, and κ, the first three of which are quite hydrophobic whereas the last is amphiphilic (Chowdhury and Luckham, 1995). A mixture of all four casein types will readily form micelles if the critical micelle concentration of 500 μg/ml is exceeded (Nylander and Wahlgren, 1994), so concentrations should be kept below this if possible. We have found that including 80 μg/ml of pure κ-casein in the experimental buffer almost completely eliminates strong, permanent adhesion of probe particles (polystyrene or silica) to surfaces without forming micelles that can interfere with data collection.

If the selected probe is to be used for topographic imaging, any interaction between probe and surface should generally be avoided. However, to explore the interaction between two molecules or between two types of surfaces, the probe must also be prepared by coating it with molecules of interest. Although molecules often adsorb tightly to colloidal particles, covalent attachment of molecules, preferably in a known orientation, is a more reliable means of probe preparation. Many protocols exist for the covalent immobilization of proteins on silica or polystyrene particles. Several companies now supply particles fabricated in a wide range of sizes and with various reactive groups on their surfaces. Perhaps the most commonly used variety are carboxylate-modified polystyrene particles. Proteins can easily be coupled to such beads via a carbodiimide-mediated reaction or via one of many available heterobifunctional cross-linkers. By choosing a longer cross-linker, such as a heterobifunctional-poly(ethylene glycol) (Haselgrübler *et al.*, 1995), the protein can be dangled further away from the bead, thereby reducing the potential for nonspecific bead–surface adsorption and also allowing the protein to assume a more natural range of orientations.

B. Scanning

As with the AFM, imaging topographic features or forces with an OFM requires a means of precisely (with subnanometer accuracy) controlling the position of the probe relative to the sample. The most straightforward way of doing this is to mount the sample of an x-y-z piezo stage, which can then be driven by a computer via a digital-to-analog converter. Such stages are commerically available, but they can be quite costly. A less expensive alternative is to construct one from piezoelectric elements. A particularly elegant design employs three pairs of piezo bimorphs, each wired to bend in an S shape (Matey *et al.*, 1987; Muralt, *et al.*, 1986). Such a stage is capable of fairly large (20–30 μm) translations at comparatively low voltage but can have a fairly low resonant frequency (150–300 Hz). For more rapid movement, it is preferable to scan the trapping beam instead of the sample. Using an acousto-optic modulator one can access frequencies in the Mhz range or higher. However, force measurements require that the final signal reflect the position of the trapped particle relative to the center of the trapping laser beam focus. Thus, scanning the beam necessitates either simultaneous scanning of the detector system as in confocal laser scanning microscopy or use of a detection scheme that does not rely on direct imaging of the trapped particle, such as the optical trapping interferometer described by Svoboda and Block (1994) after the microinterferometer development of Denk and Webb (1990).

C. Detection

As mentioned earlier, the signal of interest is the position of the probe particle relative to the minimum of the potential well formed by the optical tweezers. As the probe is moved over the sample, this position will change, due either to deflections by topographic features or to other forces acting on the probe. The probe's position is most easily detected by collecting the light of the tweezers itself, either forward or backward scattered off the trapped particle. In general, the signal from the forward-scattered light will be much larger than that from the backward-scattered light. As the position of the particle fluctuates in three dimensions, the intensity and spatial distribution of light it scatters fluctuates as well. The simplest means of detection is a focusing lens and a single photodiode placed downstream from the sample (Fig. 5A). The signal in this case is a mixture of axial and radial displacement signals. However, if the detector is calibrated by using a piezo stage to translate surface-immobilized beads through the focus of the laser beam in axial and radial directions, the relative contributions of each type of motion can be determined. Because the focal volume is elongated along the optical axis, sensitivity to motion in this direction is typically lower by a factor of 10 or so than it is to motion perpendicular to the optical axis (Ghislain *et al.*, 1994). The single photodiode configuration was used by Ghislain *et al.* (1994) for topographic imaging of polymeric grids and polystyrene beads. Sensitivity was such that features as small as 20 nm could be resolved. It should be noted that micron-scale changes in index of refraction of the sample can also contribute to the collected signal. It is therefore important to characterize the signals obtained in the absence of any trapped particle before interpreting topographic images.

Another method of detection involves imaging the laser beam focus onto the center of a quadrant photodiode. This arrangement (see Figure 5B for details) is straightforward and results in a signal with a high signal-to-noise ratio and good separation of axial from radial signals. Merely orienting the quadrant detector with its x and y axes aligned along those of the microscope stage will completely isolate x and y displacements from each other. The sum of signals from all four quadrants is relatively insensitive to radial motion perpendicular to the optical axis, a feature that makes this combination of signals convenient to use for the detection of axial (z) displacements. Although there is some cross talk between axial and radial channels, the unwanted signal tends to be less than 10% of the total signal. (Stout and Webb, unpublished data). Thus, reasonably independent measures of position fluctuations in x, y, and z directions can be obtained by using a quadrant detector.

A third variety of detector is based on the same optical principles as those in differential interference contrast (DIC) microscopy. Introduced by Denk and Webb (1990) as a method for measuring the small vibrations of hair cells, this extremely sensitive technique was later employed by Svoboda *et al.* (1993) to monitor the displacements of optically trapped silica beads as they were shuttled along microtubules by individual kinesin molecules. This method of detection requires that the trapping laser light be circularly polarized when it enters the microscope objective. A Wollaston prism before the focusing objective splits the two orthogonally polarized components of the beam and shifts them laterally with respect to each other by approximately 200 nm. This separation is small enough so that the two focused beams act as a single optical trap. A second objective, preferably identical to the first, collects the laser light after it has passed through the sample, and a second Wollason recombines the two beams. (Fig. 5C). A trapped particle centered exactly along the line between the two foci will result in light that is circularly polarized after recombination. However, if the particle is not centered, it will introduce a small degree of ellipticity into

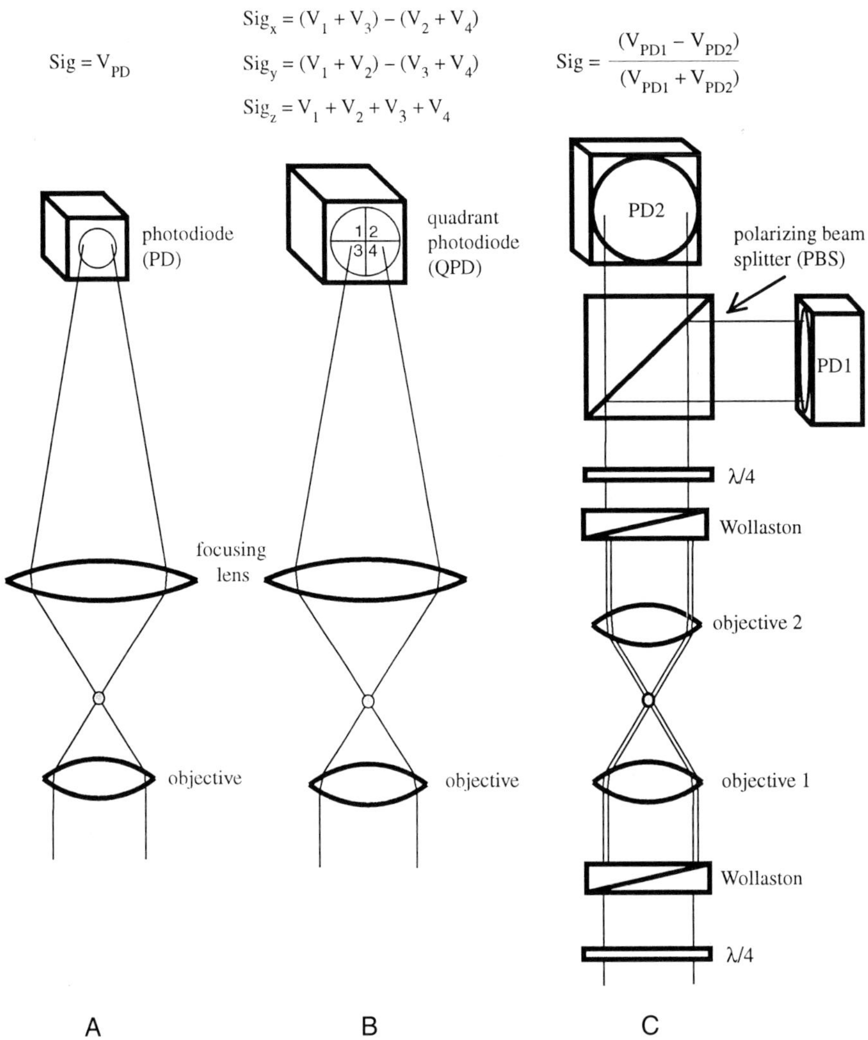

Fig. 5 Schematics of the three detection schemes described in the text, along with the signals used for each. (A) The single photodiode arrangement used by Ghislain *et al.* (1994). The photodiode is placed on the optical axis at a position that results in maximal signal amplitude. (B) The quadrant photodiode arrangement. Here the bead and optical trap are essentially imaged onto the center of the 4-diode array. A small degree of defocus may be necessary to compensate for the gap between diodes. (C) The interferometer arrangement used by Svoboda *et al.* (1993). Here linearly polarized laser light passes through a quarter-wave plate ($\lambda/4$) to become circularly polarized. Objective 1 forms the optical trap while objective 2 collects the forward-scattered light. The second $\lambda/4$ plate is not strictly necessary but does allow the user to fine-tune the zero-displacement level of the signal.

the recombined beam. A polarizing beam splitter divides the beam into two orthogonal components, the relative magnitudes of which indicate the position of the particle relative to the midpoint between the two foci. A quarter-wave plate may be placed after the second Wollaston to allow adjustment of the zero-displacement signal. This type of position detection can offer an improvement in signal-to-noise ratio over the quadrant detector because it is less sensitive to vibration (Svoboda and Bock, 1994). However, it is most effective if the motion of interest lies entirely along the line joining the two laser beam foci.

D. Calibration

To use the photodetector's voltage signals for topographic imaging or force measurements, two calibrations must be performed. First, the photodetector must be calibrated to enable conversion of voltages into actual bead displacement figures, preferably for both radial and axial displacements. It can also be useful to determine the amount of cross talk between axial and radial displacement signal channels. The most common method uses a piezoelectric translator to drive a stationary bead to known distance through the optical trap. The bead can be fixed to a substrate either by tight adsorption in moderately concentrated salt solution or by suspension in a gel (agarose or polyacrylamide) that is subsequently hardened. If the former method of fixation is used, one should note the the proximity of the glass–water interface will change the magnitude of the signal relative to what it would be for a trapped bead away from the interface. If the latter method is used, a gel as dilute as necessary should be used to firmly fix the beads because the relative index of refraction of bead and medium should be maintained as closely as possible to that of the experimental system. In either case, care must be taken to ensure that the bead is placed at the same axial plane in which it would reside when trapped by the optical tweezers because the magnitude of the forward-scattered light is highly dependent on axial position.

Second, the trap itself must be calibrated so that probe displacement can be correlated with an applied force. One means of accomplishing this is by collecting a power spectrum of position fluctuations and fitting the observed power spectral density to the expression

$$PSD(\omega) = \frac{A_0^2}{\omega^2 + \left(\frac{k}{\gamma}\right)^2}.$$

This equation give the power spectral density of position fluctuations for a particle undergoing Brownian motion in a purely harmonic potential well (see Chapter 8 for a derivation and experimental details). Variable k is the Hooke's Law constant for the optical trap along the axis of interest, and γ is the viscous drag coefficient. Although the power spectrum can provide a quick indication of trap strength, it is not the preferred method for obtaining the force–distance relationship for an optical trap. One reason for this is that often the position detector is not able completely to isolate axial from radial displacement signals so that fluctuations along two axes with two different spring constants contribute to the total power spectrum. Perhaps a more important reason is that the power spectrum is a record dominated by small displacements away from equilibrium, 5–20 nm or so. In many experiments beads are displaced much farther than this, sometimes up to 150 nm away from the trap center. Because the force-displacement curves for optical traps are frequently nonlinear for large displacements, they do not always behave like pure harmonic potentials, especially when the trapped particle is situated very close (<0.5 μm) to a glass–water interface. Therefore, the spring constant determined from a power spectrum may not provide enough information to correlate displacement with force accurately.

The most common method for calibration of the radial trap force involves generation of a fluid flow past the trapped particle by translating the sample chamber at a known velocity. The displacement of the particle away from the center of the trap is then measured for a range of forces until the applied force is great enough to push the particle out of the trap. The viscous force acting on the particle is given by the expression

$$F_{vis} = \gamma \cdot \nu,$$

where γ is the viscous drag coefficient and ν is the flow velocity. For a spherical bead, this relationship is straightforward only when the distance between the surface of the trapped sphere and a substrate is relatively large, at least four times the radius, a (Happel and Brenner, 1973). In this case, γ is given simply by the Stokes equation,

$$\gamma = 6\pi\ \eta\ a$$

where η is the solution viscosity, 0.001 kg s^{-1} m^{-1} for water. Things become more complicated as the sphere is moved closer to a surface. The preceeding expression must be corrected by a factor that depends on the sphere–surface separation. For separation of 0.02a or larger, the Faxen correction for a sphere moving through fluid parallel to single surface is sufficient (Happel and Brenner, 1973):

$$\gamma = \gamma_0 \left[1 - \left(\frac{9}{16}\right)\left(\frac{a}{h}\right) + \left(\frac{1}{8}\right)\left(\frac{a}{h}\right)^3 - \left(\frac{45}{256}\right)\left(\frac{a}{h}\right)^4 - \left(\frac{1}{16}\right)\left(\frac{a}{h}\right)^5\right]^{-1}.$$

Here h is the distance between the center of the sphere and the surface. If the gap between sphere and surface is smller than 0.02a, so that the sphere is approaching the limit of actual contact with the surface, the preceding expression is no longer accurate. In such a case, the correction of O'Neill (1964), who provided a numerical calculation of the exact solution to the limiting case in which a sphere in a viscous fluid is translated in close proximity to a planar surface, should be applied. Thus, in the case of small particle–surface separations, it is important to be able to determine the magnitude of the separation distance as closely as possible in order to get the best trap calibration.

E. Signal Processing

In addition to a sensitive detector, calibration of the optical tweezers and subsequent force measurement or imaging requires a means of acquiring, storing, and analyzing the voltage signals. Amplification of voltages from photodiodes can be carried out with simple analog circuitry and low-noise operational amplifiers. Stout and Webb (1996) utilized feedback resistors and capacitors, which result in a gain of 82 kilovolt/amp (KV/A) and a 3_{dB} point of 10 kHz for their quadrant photodiode. Both imaging and force measurements require a computer equipped with a sufficiently fast data-acquisition board that can record rapid (~5–10 kHz) changes in probe position. The capacity for capturing multiple channels of data at a rate of 10 kHz or more is essential when more than one photodetector is used for the signal. Applications requiring feedback, such as in the OFM, must consider not only the data acquisition rate but the time delay between input and output of signals. This delay should be minimized so as to introduce as little phase shift as possible between signals in and out. For force measurement applications, a system that will acquire the voltages from the stage driver and photodetectors and display them both on the same time base is the sufficient. However, for imaging applications, a more sophisticated display system is needed. Commercially available software designed for use with the AFM or near-field scanning microscopy (NSOM) is quite suited for use with the optical force microscope as well. Ghislain (1994) utilized the software supplied with the Nanoscope III AFM with no modifications.

IV. New Directions

Optical tweezers may not approach the AFM in spatial resolution or sheer strength, but they can explore realms not accessible to the AFM. One promising area is that of intracellular particle trapping. Any nucleated cell contains a myriad of vesicles and organelles, all existing within the complex matrix of the cytoplasm. Optical tweezers have been used to restrain the directed transport of mitochondria along microtubules in the giant amoeba, providing a means of estimating the force exerted by the motor molecules causing the motion (Ashkin *et al.*, 1990). More precise investigations of such a system using a high-sensitivity position sensor would enable the construction of a time-dependent picture of the force exerted, perphaps leading to an estimate of the number of motor molecules responsible for transport of one mitochondrion.

Single-particle tracking experiments have shown that cell surface proteins move about the membrane in several different modes: free diffusion, diffusion between barriers, and directed, unidirectional motion (Feder *et al.*, 1996; Ghosh and Webb, 1994). Although much work has been done to follow surface-bound particles on cells, relatively little has been done to measure the actual forces that drive or restrict their movements. Edidin *et al.* (1991) used optical tweezers to rapidly pull surface protein-bound gold microspheres along the surfaces of murine HEPA-OVA cells to determine the location and size of barrier-free regions on cell membranes. A different approach to investigating the same problem has also been initiated (Switz *et al.*, 1996). In this system, an optical trapping microscope equipped with feedback electronics provides a passive/active means of tracking the motion of a single surface-linked particle while simultaneously probing the viscosity of the surrounding plasma membrane matrix and measuring and drag force fluctuations and active driving forces that arise from within the cell.

Once fully functional, such an apparatus could also be used to follow the motions of intracellular particles. Observations of tracer particles microinjected into living cells have shown that the cytoplasm is by no means homogeneous but is made up of domains of varying size (Luby-Phelps *et al.*, 1987). The optical tweezers are capable of holding and manipulating intracellular vesicles inside cells without damaging them. Measurements of heating at the focus of an optical trap have shown that the local temperature rise is on the order of 1°C (Liu *et al.*, 1994), an amount that should not damage cells if they are maintained at 25–35°C during experiments. In a manner akin to the cell surface experiments described earlier, the feedback-modulated optical trap might be used to probe barriers and measure the viscoelastic properties of intracellular domains using either native or microinjected particles.

One area that has received surprisingly little attention is the potential for multiphoton excitation of fluorescence at the focus of an optical trap. Although the lasers used for optical tweezers are typically continuous-wave lasers, with power at the focus ranging from 10 to approximately 100 mW, the intensity at the focus of a high numerical aperture (NA = 1.3) can approach 20 mW/cm^2. Such a high-photon flux density should be capable of exciting two-photon fluorescence, at least for fluorophores with a reasonably high excitation cross-section at the trapping wavelength. Indeed, when orange fluorescent microspheres are held in a 1064-nm continuous-wave laser trap with incident power of approximately 85 mW, fluorescence can be observed in the absence of any other illumination source (data not shown).

Pulsed lasers are also capable of generating optical forces suitable for trapping small particles. Several experiments have been performed on suspensions of either 30- or 500-nm blue fluorescent beads utilizing a Ti:sapphire laser beam (λ = 800 nm), which provided an average power of less than 100 mW at the sample in an 80-MHz pulse train with 250-femtosecond pulse, width. The laser beam was focused into the sample via a 1.2-NA water-immersion objective in these investigations. (J. Mertz, C. Xu, and W. Webb, unpublished data). Both types of beads emitted bright two-photon-excited fluorescence while being stably trapped. The smaller beads photobleached after approximately 0.5 sec, whereas the larger beads exhibited only slow photobleaching. Switching the laser between pseudocontinuous wave operation and mode-locked pulsed operation allowed benign trapping in the continuous-wave mode with brief test periods of pulsed mode to probe particle fluorescence and position. Such preliminary results are a promising indication that it should be possible to perform multiphoton-excited fluorescence imaging of intra- or extracellular particles (native vesicles or introduced microparticles) while simultaneously manipulating them with optical tweezers.

References

Ashkin, A., Schütze, K., Dziedzic, J. M., Euteneuer, U., and Schliwa, M. (1990). Force generation of organelle transport measured *in vivo* by an infrared laser trap. *Nature* **428,** 346–348.

Berger, C. E. H., van der Werf, K. O., Kooyman, R. P. H., de Grooth, B. G., and Greve, J. (1995). Functional group imaging by adhesion AFM applied to lipid monolayers. *Langmuir* **11,** 4188–4192.

Boland, T., and Ratner, B. D. (1995). Direct measurement of hydrogen bonding in DNA nucleotide bases by antomic force microscopy. *Proc. Natl. Acad. Sci. USA* **92,** 5297–5301.

Chowdhury, P. B., and Luckham, P. F. (1995). Interaction forces between κ-casein adsorbed on mica. *Colloids and Surfaces B* **4,** 327–334.

Dabros, T., Warszynski, P., and van den Ven, T. G. M. (1994). Motion of latex spheres tethered to a surface. *J. Coll. Interface Sci.* **162,** 254–256.

Denk, W., and Webb, W. W. (1990). Optical measurement of picometer displacements of transparent microscopic objects. *Appl. Optics* **29,** 2382–2390.

Edidin, M., Kuo, S. C., and Sheetz, M. P. (1991). Lateral movements of membrane glycoproteins restricted by dynamic cytoplasmic barriers. *Science* **254,** 1379–1382.

Feder, T. J., Brust-Mascher, I., Slattery, J. P., Baird, B., and Webb, W. W. (1996). Constrained diffusion or immobile fraction on cell surfaces: A new interpretation. *Biophys. J.* **70,** 2767–2773.

Felgner, H., Müller, O., and Schliwa, M. (1995). Calibration of light forces in optical tweezers. *Appl. Optics* **34,** 977–982.

Finer, J. T., Simmons, R. M., and Spudich, J. A. (1994). Single myosin molecule mechanics: Piconewton forces and nanometre steps. *Nature* **368,** 113–119.

Florin, E-L., Moy, V. T., and Gaub, H. E. (1994). Adhesion forces between individual ligand-receptor pairs. *Science* **264,** 415–417.

Frisbie, C. D., Rozsnyai, L. F., Noy, A., Wrighton, M. S., and Lieber, C. M. (1994). Functional group imaging by chemical force microscopy. *Science* **265,** 2071–2074.

Ghislain, L. P. (1994). "Force microscopy based on the optical trapping of microscopic particles." Ithaca, New York: Cornell University. Ph.D. thesis.

Ghislain, L. P., and Webb, W. W. (1993). Scanning-force microscope based on an optical trap. *Opt. Lett.* **18,** 1678–1680.

Ghislain, L. P., Switz, N. A., and Webb, W. W. (1994). Measurement of small forces using an optical trap. *Rev. Sci. Instrum.* **65,** 2762–2768.

Happel, J., and Brenner, H. (1973). "Low Reynolds number hydrodynamics with special applications to particulate media." 2d ed., p. 327. Leyden, The Netherlands: Noordhoff International Publishing.

Haselgrübler, T., Amerstorfer, A., Schindler, H., and Gruber, H. J. (1995). Synthesis and applications of a new poly(ethylene glycol) derivative for the crosslinking of amines with thiols. *Bioconjugate Chem.* **6,** 242–248.

Hinterdorfer, P., Baumgartner, W., Gruber, H. J., Schilcher, K., and Schindler, H. (1996). Detection and localization of individual antibody–antigen recognition events by atomic force microscopy. *Proc. Natl. Acad. Sci. USA* **93,** 3477–3481.

Israelachvili, J. (1992). "Intermolecular and surface forces." 2d ed. Chapter 12. San Diego.: Academic Press.

Lal, R., and John, S. A. (1994). Biological applications of atomic force microscopy. *Am. J. Physiol.* **266** (*Cell Physiol.* **35**), C1-C-21.

Lee, G. U., Chrisey, L. A., and Colton, R. J. (1994). Direct measurement of the forces between complementary strands of DNA. *Science* **266,** 771–773.

Liu, Y., Cheng, K., Sonek, G. J., Berns, M. W., and Tromberg, B. J. (1994). Microfluorometric technique for the determination of localized heating in organic particles. *Appl. Phys. Lett.* **65,** 919–921.

Luby-Phelps, K., Castle, P. E., Taylor, D. L., and Lanni, F. (1987). Hindered diffusion of inert tracer particles in the cytoplasm of mouse 3T3 cells. *Proc. Natl. Acad. Sci. USA* **84,** 4910–4913.

Matey, J. R., Crandall, R. S., Brycki, B., and Briggs, G. A. D. (1987). Bimorph-driven x-y-z translation stage for scanned image microscopy. *Rev. Sci. Instrum.* **58,** 567–570.

Muralt, P., Pohl, D. W., and Denk, W. (1986). Wide-range, low-operating-voltage bimorph STM: Application as potentiometer. *IBM J. Res. Dev.* **30,** 443–450.

Nylander, T., and Wahlgren, N. M. (1994). Competitive and sequential adsorption of β-casein and β-lactoglobulin on hydrophobic surfaces and the interfacial structure of β-casein. *J. Coll. Interface Sci.* **162,** 151–162.

O'Neill, M. E. (1964). A slow motion of viscous liquid caused by a slowly moving solid sphere. *Mathematika* **11,** 67–74.

Stout, A. L., and Webb, W. W. (1998). Measuring intermolecular forces with a scanning optical trap. *Biophys. J.* Submitted.

Svoboda, K., Schmidt, C. F., Schnapp, B. J., and Block, S. M. (1993). Direct observation of kinesin stepping by optical trapping interferometry. *Nature* **365,** 721–727.

Svoboda, K., and Block, S. M. (1994). Biological application of optical forces. *Annu. Rev. Biophys. Biomol. Struct.* **23,** 47–85.

Switz, N. A., Mertz, J., and Webb, W. W. (1996). A feedback modified optical trap for probing local viscosity and examining diffusive behavior on cell membranes. *Biophys. J.* **70,** A334.

Visscher, K., and Brakenhoff, G. J. (1992a). Theoretical study of optically induced forces on spherical particles in a single beam trap I. Rayleigh scatterers. *Optik* **89,** 174–180.

Visscher, K., and Brakenhoff, G. J. (1992b). Theoretical study of optically induced forces on spherical particles in a single beam trap II. Mie scatterers. *Optik* **90,** 57–60.

Wohland, T., Rosin, A., and Stelzer, E. H. K. (1996). Theoretical determination of the influence of the polarization on forces exerted by optical tweezers. *Optik* **102,** 181–190.

Wright, W. H., Sonek, G. J., and Berns, M. W. (1994). Parametric study of the forces on microspheres held by optical tweezers. *Appl. Optics* **33,** 1735–1748.

Yin, H., Wang, M. D., Svoboda, K., Landick, R., Block, S. M., and Gelles, J. (1995). Transcription against an applied force. *Science* **270,** 1653–1657.

CHAPTER 7

Single Molecule Imaging and Nanomanipulation of Biomolecules

Yoshie Harada,[*] Takashi Funatsu,[*] Makio Tokunaga,[*] Kiwamu Saito,[*] Hideo Higuchi,[*] Yoshiharu Ishii[*] and Toshio Yanagida,[*†]

[*] Yanagida BioMotron Project, ERATO, JST
Senba-higasi 2-4-14, Mino
Osaka, 562 Japan

[†] Department of Physiology
Osaka University Medical School, Suita, Osaka 565 Japan

I. Introduction

The best way to provide unambiguous information about the working principle of biomolecular machines such as molecular motors is to directly observe and manipulate individual molecular machines under an optical microscope. Because of the diffraction limit, however, the resolution of conventional optical microscopes (~0.2 μm) is too low for observing single protein molecules, or even their

Copyright © 1998 by Academic Press. All rights of reproduction in any form reserved.
0091-679X/98 $25.00

small assembly. Thus, a great deal of effort has been made to increase the resolution of the experimental system. One method is to observe the scattered light from objects by using a dark-field microscope equipped with a very strong illuminating source and a highly sensitive camera. However, the objects observed need to be as large as a microtubule (~20 nm) in diameter (Hotani, 1976). A similar resolution can be achieved if the contrast of the recording is greatly enhanced by computer image processing (video-enhanced contrast method) (Inoue, 1981).

Fluorescence microscopy is useful for observing smaller objects. About 10 years ago, it was demonstrated that single actin filaments (double-helical polymer of actin), labeled with fluorescent phalloidin, can be clearly seen by fluorescence microscopy (Yanagida *et al.,* 1984). This finding led to the development of new *in vitro* motility assays that address the elementary process of force generation by actin–myosin interaction directly at the molecular level (Harada *et al.,* 1987; Kishino and Yanagida, 1988; Kron and Spudich, 1986; Toyoshima *et al.,* 1987). More recently, combining of techniques for manipulation of an actin filament and nanometry has allowed the individual mechanical events driven by adenosine 5′-triphosphate (ATP) hydrolysis to be measured at sub-piconewton and sub-nanometer resolutions directly from multiple (Finer *et al.,* 1994) and single myosin molecules (Ishijima *et al.,* 1991; Ishijima *et al.,* 1994).

The number of photons emitted from a single fluorophore excited by a strong light source such as a laser is sufficient for it to be visualized by a commercial high-sensitivity video camera. Thus, it is theoretically possible to visualize single biomolecules labeled with a fluorescent dye, but until recently it was not yet proven in practice owing to great background noise. Recently, new fluorescence microscopes have been developed with 20- to more than 2000-fold less background noise than a conventional fluorescence microscope and thus enable visualizing single fluorophores in aqueous solution (Fanatsu *et al.,* 1995). Here, we demonstrate that the movements and individual ATP turnovers of single motor proteins can be directly visualized at the real time.

II. Visualization of Single Fluorophores in Aqueous Solution

When a laser beam of 514-nm wavelength and 10-mW power is focused to 40-μm diameter at the object, the photon density is $\sim 10^{13}$ photons/sec/μm^2. Assuming the absorption coefficient of fluorophores ($10^5 M^{-1} cm^{-1}$), a quantum yield of 0.5, and that 40% of the photons emitted from the fluorophores reach the camera through a objective lens, the number of photons available for imaging would be approximately 10^5 photons/sec/fluorophore. If the image is projected on the camera with the magnification of 1000$\times$ and the size of the image of a single fluorophore corresponds to the diffraction limit area (0.25×0.25 μm at the objective or 250×250 μm at the camera), the photon density at the camera would be approximately 10^8 photons/sec/cm^2 (Harada and Yanagida, 1988). This photon density is much larger ($\sim 10^5$/sec/cm^2) than required for observation by the silicon intensified target (SIT) camera, the most widely used high-sensitivity instrument. The problem, however, is that the background luminescence and Raman scattering by water are usually huge. For example, in a conventional fluorescence microscope, the background is more than 10 times larger than the fluorescence from a single fluorophore. Therefore, the key point for imaging single fluorescent molecules is how to reject the background.

A. Low-Background Epifluorescence Microscopy (LBEFM)

We tried to minimize the background in a conventional epifluorescence microscope. Origins of the background were luminescence from an objective lens, immersion oil, a coverslip, dusts in solution on the surface of a coverslip, Raman scattering from immersion oil and solution, and incident light breaking through filters. We replaced an objective lens by a carefully chosen nonfluorescence lens, immersion oil by mixture of two kinds of silicone oil, and a coverslip by a quartz one. Fluorescent dye (cyanine dyes Cy3 and Cy5), sharp-cut barrier filters, and the wavelength of incident light (laser) were appropriately chosen to minimize contamination of Raman scattering and incident light. Dusts, especially those attached to the surface of a coverslip, were very difficult to remove because the surface was immediately blotted in air. Therefore, coverslips were cleaned and experiments were conducted done in a clean room. Dust in solution was removed by filtration. Thus, the background was reduced from more than 6000 to 140 photons/sec/diffraction limit area ($0.25 \times 0.25\ \mu m$) at the power of 10 mW of an incident argon laser.

Fig. 1A shows fluorescent images of myosin subfragments: heavy meromyosin (HMM) labeled with fluorescent dye molecules (Cy3). The dye molecules bound specifically to a reactive thiol group (SH-1) on each head of HMM, which has two heads. Because the average molar ratio of dye to head is approximately 1, some HMM molecules (~50%) are expected to bind two dyes, with the rest binding one dye (~40%) or no dye (~10%). Fig. 1B shows the electron micrograph of the same field. Single HMM molecules, identified by their characteristic two-headed shape, were found in positions corresponding to the least intense fluorescent spots, or in positions of more intense fluorescent spots, of which intensity was approximately two times larger than that of the least intense spots (the second Gaussian peak in Fig. 2). Fig. 2 shows the intensity histograms of individual fluorescent spots. Fluorescent spot intensities were quantized at intervals of approximately 500 photons/sec. Fluorescence intensities of more than 1500 photons/sec very likely result from two HMM molecules located in close proximity. Taken together, the evidence described provides strong support for the assertion that single fluorescent molecules can be visualized by LBEFM (Funatsu *et al.,* 1995).

B. Low-Background Total Internal Reflection Fluorescence Microscopy (LBTIRFM)

LBEFM is capable of visualizing single fluorophores in solution but not sufficient for clearly observing single fluorophores at a full-video rate or better

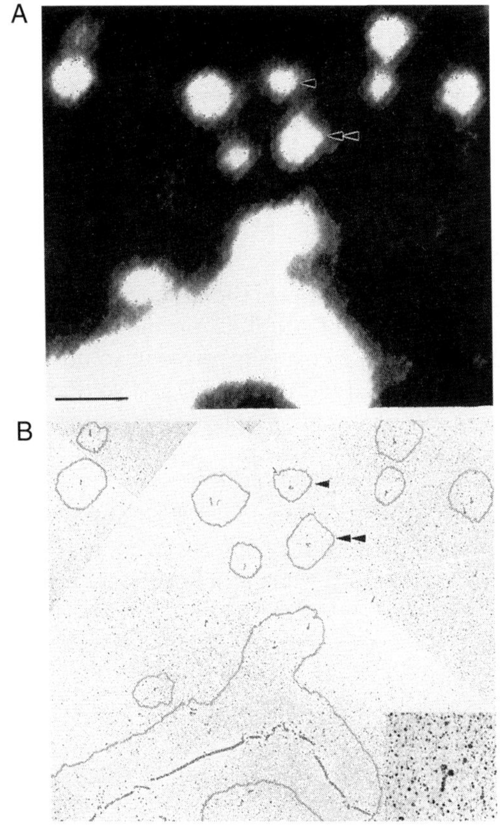

Fig. 1 Low-background epifluorescence microscopy (LBEFM, Funatsu, *et al.*, 1995). (A) Fluorescence micrograph taken by LBEFM. Cy3-labeled HMMs were spread on a thin mica film to enable subsequent electron microscopy and illuminated by argon laser. The fluorescence (λ_{em} = 555–585 nm) images were videotaped with I ICCD camera and 64 frames were averaged. As a marker to identify position, phalloidin tetramethyl rhodamine-labeled actin filaments were introduced as seen in the lower left. The typical fluorescent spots due to one and two dye molecules are indicated by single and double arrowheads, respectively. Bar, 1 μm. *Optics:* An inverted microscope equipped with epifluorescence optics (TMD300, Nikon, Inc., Japan) was modified as follows. A beam of an argon laser (514.5 nm) or a helium–neon laser (632.8 nm) was depolarized by passage through a 1/4 plate and focused by lens on a rotating ground glass to eliminate inhomogeneity of illumination due to interference. The depolarized and random-phase beam was passed through a pinhole (P) and was collimated with lens. Then the beam was focused on an aperture diaphragm and introduced to the nonfluorescent objective lens (Plan Apo 100×; 1.4 NA) through a dichroic mirror (made specially to reflect 99.9% of the incident beam). The specimens were excited by paraxial rays with Koehler illumination. All glass optical parts, except for the objective lens, were replaced with those made of fused quartz, which emit little fluorescence. Nonfluorescent silicone oil was used as immersion oil. Background luminescence was rejected with a barrier filter. To reduce dust contamination, which was a serious source of background noise, coverslips and slides were washed with 0.1 *M* KOH and EtOH, then dried in a class 1000 clean room. Experiments were also performed in the clean room. Dust in solution was removed by filtration before experiments. (B) Rotary-shadowed electron micro-

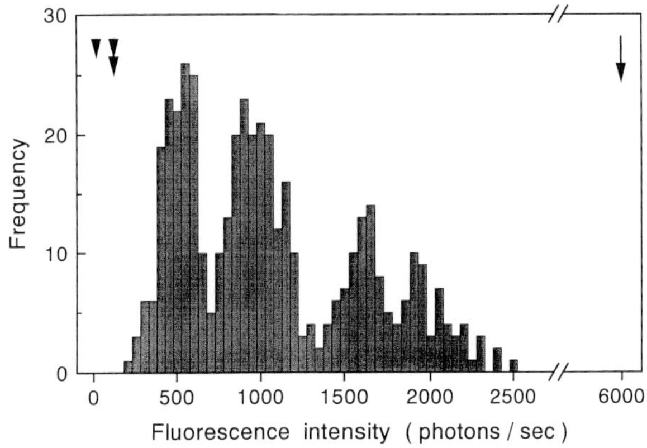

Fig. 2 The histogram of fluorescence intensity from Cy3-labeled HMMs. The laser power was 10 mW at the objective plane, and the illumination area was 150 μm in diameter. Images were taken with a cooled CCD at 5-sec exposure. The number of photons that arrived at the camera was calculated from the number of photoelectrons and quantum efficiency (~30%) of the camera. The double and single arrow heads indicate the background in LBEFM and LBTIRFM, respectively. The arrow indicates the background before refinement.

because the background is still high. To further decrease the background luminescence, we reduced the optical excitation volume using evanescent field illumination. When a laser was incident on a quartz slide at greater than the critical angle and totally reflected at the interface between the quartz and solution, the evanescent field is produced just beyond the interface (Fig. 3A). This evanescent field was localized near the interface, and the 1/e penetration depth was 100 to 200 nm, depending on the incident angle of the laser. Combining the low-background optics with the local-excitation optics, the background was furthermore reduced to 1 to 3 photons/sec/diffraction limit area (i.e., >2000 times lower than that of a conventional epifluorescence microscope). Thus, single fluorophores bound to HMM molecules could be visualized clearly at a full-video rate (1/30 sec) without frame averaging (Fig. 3B, Funatsu *et al.,* 1995). The high-rate imaging allowed

graph of the same field and magnification as those in A. HMM molecules are colored red. The inset shows the high-magnification image of an HMM. The same actin filament seen by video is also seen at lower left. The outlines of the fluorescence images are shown by green lines. Some HMM molecules were located close to one another and generally corresponded to intense spots. The population of the nonfluorescent HMMs (approximately 30%) was slightly higher than that estimated from the labeling ratio (10%). This is probably because some dyes were bleached before labeling reaction or during the focusing of an objective lens on them.

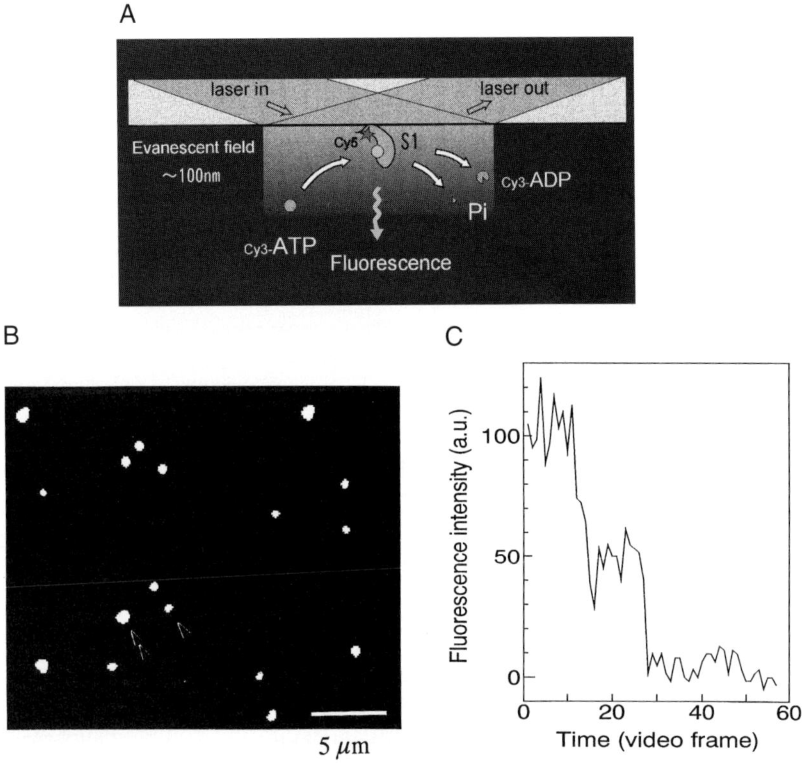

Fig. 3 Low-background total internal reflection fluorescence microscopy (LBTIRFM, Funatsu *et al.*, 1995) (A) The schematic drawing showing the principle of single molecule imaging. This is for imaging fluorescently labeled S1 molecules bound to the glass surface as well as fluorescent ATP analogue, Cy3-ATP, bound to the S1 (see text and Fig. 5). The laser beam was incident on a quartz microscope slide through a 60° dispersion prism. The gap between the microscope slide and prism was filled with nonfluorescent pure glycerol. Incident angle at the quartz slide-to-solution interface was 68° to the normal (the critical angle of 65.5°). The beam was focused by a lens to be 100 × 200 μm at the specimen plane. A cooled CCD camera was used for quantitative analysis of the fluorescence intensity with low temporal resolution. For observation of rapid movement or changes in fluorescence intensity, an I ICCD camera or ISIT camera was used. (B) The micrograph shows one frame image of Cy3-labeled HMMs taken at the video rate (exposure time, 1/30 sec) without averaging. The laser power was 15 mW, at which the average lifetime of fluorophores was approximately 15 sec. Single and double arrowheads indicate typical fluorescent spots due to one or two dye molecules bound to HMM, respectively. Lower intense spots are also seen, but due to shot noise of the I ICCD camera, which disappeared in frame-averaged images. Bar, 5 μm. (C) Quantized photobleaching of fluorescent molecules observed at the video rate (1/30 sec).

us to see clearly the time course of stepwise photobleaching of fluorescence from two dye molecules, which probably were bound to a single HMM (Fig. 3C).

III. Application

A. Direct Observation of Single Kinesin Molecules Moving along Microtubules

We applied the LBTIRFM to observe movements of single kinesin molecules along a microtubule (Vale *et al.*, 1996). Kinesin is composed of two heavy chains, each consisting of an N-terminal force-generating domain, a long α-helical coiled coil, and a small globular C-terminal domain that may bind to organelles. To fluorescently label kinesin without losing the function of the motor, the kinesin gene was truncated near the center of the α-helical coiled-coil region at amino acid 560. Then a short peptide sequence containing a highly reactive cysteine was introduced at the C terminus. The truncated derivatives were expressed in bacteria and reacted at the C-terminal cysteine residue with a maleimide-modified Cy3 fluorescent dye.

Cy3-labeled kinesin derivatives (0.6 n*M*) were applied to a Cy5-labeled flagellar axoneme (a 9 + 2 microtubule array) adsorbed on the surface of a quartz slide, which had been previously observed and illuminated by the surface evanescent field (Fig. 3A). Kinesins associated with the axoneme could be seen as clear, in-focus spots, whereas those undergoing random thermal motion moved too rapidly to be detected as discrete intensities and only contributed to a diffuse background in the image. Fig. 4 shows sequential fluorescence images of a single kinesin molecule moving along an naxoneme. The velocity of fluorescent kinesin movement was approximately 0.3 μm/sec, which is similar to the velocity of axoneme moving over a glass surface coated with the kinesin derivatives.

Thus, the movements of single fluorescently labeled motor proteins can be clearly seen.

B. Direct Observation of Individual ATP Turnovers by Single Myosin Molecules

Single S1 molecules (single-headed myosin subfragments) were labeled with Cy5 and fixed on the quartz slide surface. The ATP turnover events were detected by directly observing association-(hydrolysis)-dissociation of fluorescent ATP analog labeled with Cy3. When 10-n*M* Cy3-ATP was applied to S1 on the surface, the background fluorescence due to free Cy3-ATP was low because the illumination region was localized near the quartz slide surface (Fig. 3A). When Cy3-ATP or -ADP was associated with surface-bound Cy5-S1 that had been previously visualized (Fig. 5A), it could be seen as a clear, in-focus fluorescent spot (Fig. 5B). In contrast, free Cy3-ATP undergoing rapid Brownian motion was not seen as a discrete spot. Hence, by observing the presence and lifetime of stationary, in-focus Cy3 molecules corresponding to the position of Cy5 S1

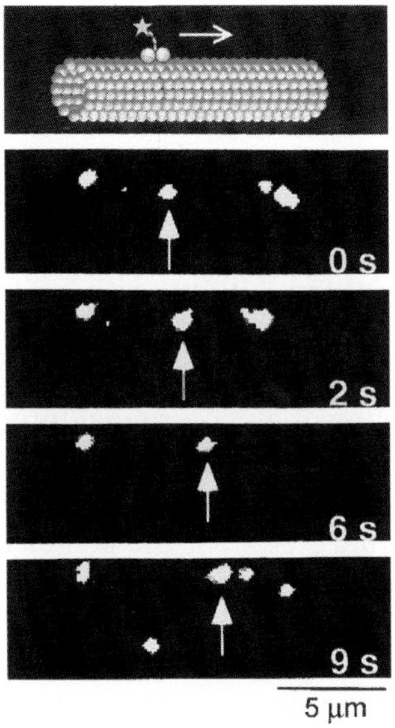

Fig. 4 Movement of a single fluorescently labeled kinesin molecule along a microtubule observed by LBTIRFM (Vale *et al.*, 1996). Upper panel shows schematic diagram. Lower three panels show sequential images of a single kinesin molecule moving along a microtubule. Times at which micrographs are taken are indicated in each panel.

molecules on the surface, individual association-(hydrolysis)-dissociation of Cy3-ATP with single S1 could be detected (Fig. 5A). Fig. 5B shows the association and dissociation of single Cy3-ATP/ADP molecules with a single S1 molecule (indicated by an arrow head in Fig. 5A). Fig. 5C shows the lifetime histogram of bound Cy3-ATP or -ADP, which reveals an exponential dissociation rate $k_{-} = 0.059\ \text{sec}^{-1}$. The photobleaching of Cy3-ATP hardly affects the results because its lifetime when bound to a quartz surface is approximately 85 sec at the same laser power (3.8 mW). The dissociation rate is in good agreement with the ATP turnover rate of Cy3-ATP by S-1 suspended in solution ($0.045 \pm 0.002\ \text{sec}^{-1}$), suggesting that the binding and dissociation of Cy3-ATP or -ADP visualized by microscopy indeed reflects a single ATP turnover. Because individual S1 molecules could turn over Cy3-ATP during several minutes of observation, illumination of Cy3-nucleotide bound to S1 does not appear to diminish enzymatic activity. Here, very slow ATP turnover events are shown. However, if the fluorescence intensity from spots due to bound fluorescent nucleotides is measured by a high-sensitivity detector such as a photon-counting detector, much faster events in the millisecond range can be detected as shown later.

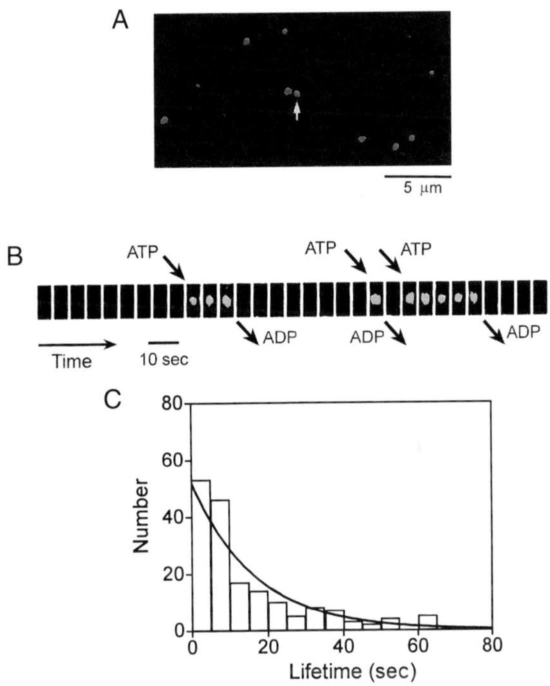

Fig. 5 Visualization of individual ATP turnovers by single S1 molecules (Funatsu *et al.*, 1995). (A) Fluorescence micrograph of single Cy5-labeled S1 molecules bound to the surface. Cy5-labeled S1 molecules were illuminated with a helium–neon laser of 5 mW and emitted red (~670 nm) fluorescence. The images were artificially colored red. Bar, 5 μm (B) ATP turnovers by a single S1 molecule. Upper panels show typical images of fluorescence from Cy3-nucleotide (ATP or ADP) coming in and out of focus by associating and dissociating with a S1 indicated by the arrowhead. Bound Cy3-nucleotides were illuminated with an argon laser of 3.8 mW and emitted yellow (~570 nm) fluorescence. The images were artificially colored yellow. Lower trace, time course of the corresponding fluorescence intensity. (C) Histogram of the lifetime of Cy3-nucleotides bound to S1s. The lifetime of bound Cy3-nucleotides was determined by measuring durations while Cy3-nucleotides bound to S1s made clear fluorescent spots as shown in B. The dissociation rate (1/the lifetime) was determined to be 0.059 sec^{-1} from the linear regression of the logarithmic plot, which was consistent with that determined using an S1 suspension.

C. Individual ATP Turnovers by a Single Kinesin Molecule Manipulated by Optical Tweezers

The schematic drawing in Fig. 6A shows a principle of measurements of individual ATP turnovers by a single kinesin molecule trapped by optical tweezers. LBTIRFM for single molecule imaging was combined with that for optical tweezers (refer to Chapter 4). A single kinesin molecule was attached to a polystyrene bead of 1-μm diameter trapped by an infrared laser (Svoboda *et al.*,

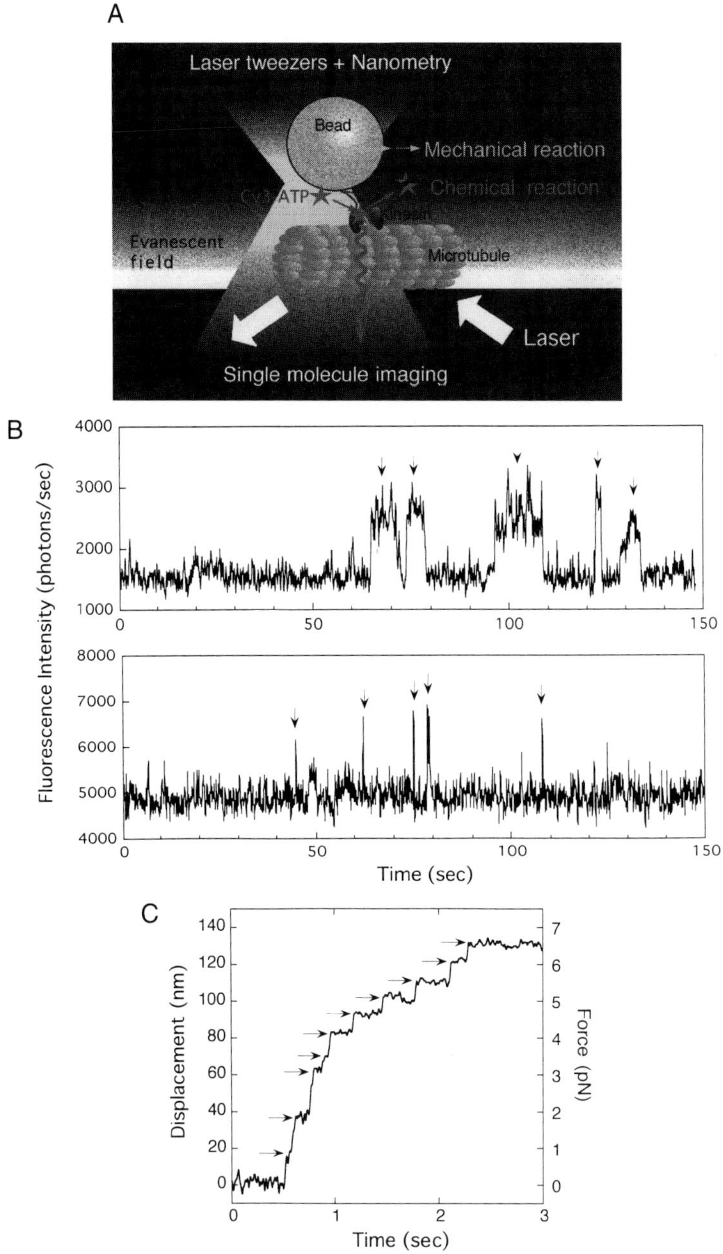

Fig. 6 Measurements of the mechanical elementary events and individual ATP turnovers by a single kinesin molecule. (A) Schematic drawing of the principle of measurements. Optics for single molecule imaging, optical tweezers, and nanometer sensing are combined. (B) Association and dissociation of Cy3-ATP or -ADP with a single kinesin molecule. Fluorescence intensities from associated Cy3-nucleotide were measured by an avalanche photodiode. Upper and lower traces indicate the events when a kinesin molecule associated with and dissociated from a microtubule, respectively. The concentration of Cy3-ATP is 50 nM. (C) Mechanical elementary events (i.e., 8-nm steps by a single kinesin molecule detected by the system shown in A).

1993). By controlling the position of a bead by the optical tweezers, the kinesin molecule was brought into contact with a microtubule adsorbed onto the glass surface. The individual ATP turnover events caused by the kinesin were measured by using the fluorescent ATP analog, Cy3-ATP, as shown earlier. Fig. 6B shows the time course of the fluorescence intensities from Cy3-ATP or -ADP bound to the kinesin measured by a photon-counting detector (i.e., individual ATP turnovers). The frequency of turnover events was rather small because the concentration of Cy3-ATP added was very low (50 n*M*). When the kinesin molecule was detached from the microtubule, the lifetime of bound Cy3-nucleotide was about 10 sec (upper trace), and the lifetime when attached (0.08 sec) was greatly shortened (lower trace). Because the lifetime of bound nucleotides is nearly equal to the ATP turnover time at saturate ATP concentration, the result shows that the ATPase activity of the kinesin is greatly activated by the microtubule as in solution.

Furthermore, because the present system is equipped with a nanometer sensor (Finer *et al.,* 1994; Ishijima, *et al.,* 1991; Svoboda, *et al.,* 1993), the elementary mechanical events (i.e., 8-nm steps of a single kinesin molecule) also can be measured (Fig. 6C). Thus, it is now possible to simultaneously measure the individual ATP turnovers and the elementary mechanical events of a single kinesin molecule (Funatsu *et al.,* 1996). This should provide a clear answer to the fundamental problem of how the mechanical reaction is coupled to the ATPase reaction. The results will be published elsewhere in the near future.

IV. Perspectives

The methods described here also can be applied to examining single nucleotidase reactions of other enzymes (e.g., interactions of polymerases and helicases with DNA). It is possible to suspend a single DNA in solution by manipulation with dual optical traps, as well as to see single fluorescently labeled RNA polymerases. Directly observing the reading process of the DNA genetic information by a single RNA polymerase molecule is not just a dream now, but realistic. Furthermore, the single molecule imaging method enables imaging of fluorescence energy resonance transfer between a single donor and a single acceptor bound to biomolecule(s): single molecule fluorescence energy resonance transfer (SFERT) and single molecule spectroscopy (SMP) (Ishii *et al.,* 1997). The SFRET and SMP will enable use to examine conformational states of individual protein molecules and follow the protein-folding as well as the protein–protein or protein–ligand association process at a single molecular level.

Thus, the techniques for single molecule imaging and manipulation will be very powerful for studies not only of motility of motor proteins, but also of molecular genetics, signal transduction and processing in cell, dynamic molecular process of proteins.

References

Finer, J. T., Simmons, R. M., and Spudich, J. A. (1994). *Nature* **368,** 113–119.

Funatsu, T., Harada, Y., Tokunaga, M., Saito, K., and Yanagida, T. (1995). *Nature* **374,** 555–559.

Funatsu, T., Harada, Y., Higuchi, H., Tokunaga, M., Saito, K., Vale, R. D., and Yanagida, T. (1996). *Biophys. J.* **70,** A6.

Harada, Y., Noguchi, A., Kishino, A., and Yanagida, T. (1987). *Nature* **326,** 805–808.

Harada, Y., and Yanagida, T. (1988). *Cell Motil. Cytoskeleton* **10,** 71–76.

Hotani, H. (1976). *J. Mol. Biol.* **106,** 151–166.

Inoue, S. (1981). *J. Cell Biol.* **89,** 346–356.

Ishii, Y., Funatsu, F., Wazawa, T., Yoshida, T., Watai, J., Ishii, M., and Yanagida, T. (1997). *Biophys. J.* **72,** A283.

Ishijima, A., Doi, T., Sakurada, K., and Yanagida, T. (1991). *Nature* **352,** 301–306.

Ishijima, A., Harada, Y., Kojima, H., Funatsu, T., Higuchi, H., and Yanagida, T. (1994). *Biochem. Biophys. Res. Commun.* **199,** 1057–1063.

Kishino, A., and Yanagida, T. (1988). *Nature* **334,** 74–76.

Kron, S. J., and Spudich, J. A. (1986). *Proc. Natl. Acad. Sci. USA* **83,** 6272–6276.

Svoboda, K., Schmidt, C. F., Schnapp, B. J., and Block, S. M. (1993). *Nature* **365,** 721–727.

Toyoshima, Y. Y., Kron, S. J., McNally, E. M., Niebling, K. R., Toyoshima, C., and Spudich, J. A. (1987). *Nature* **328,** 536–539.

Vale, R. D., Funatsu, T., Romberg, L., Pierce, D. W., Harada, Y., and Yanagida, T. (1996). *Nature* **380,** 451–453.

Yanagida, T., Nakase, M., Nishiyama, K., and Oosawa, F. (1984). *Nature* **307,** 58–60.

CHAPTER 8

Signals and Noise in Micromechanical Measurements

Frederick Gittes and Christoph F. Schmidt

Department of Physics, and Biophysics Research Division
University of Michigan
930 North University
Ann Arbor, Michigan 48109

I. Introduction

A great deal is known about the static structure of the most important building blocks of life—proteins and nucleic acids—but relatively little about their motions. Intramolecular motions are, however, a central feature of the biological function of biomolecules. Thus there is great potential in new techniques that

Copyright © 1998 by Academic Press. All rights of reproduction in any form reserved.
0091-679X/98 $25.00

make it possible to study the dynamics of individual biological macromolecules. A variety of single-molecule experiments, ranging from optical tweezers and scanned-tip microscopies to single-molecule fluorescence methods, have recently begun to explore the new territory. Researchers are faced with a multitude of challenging problems, one of which is noise that sets limits on the resolution of single-molecule measurement. Instrumentation must be designed with enough stability to make measurements on nm-length scales, and a thorough understanding of the subtleties of data analysis is necessary to push the limits of detection and to avoid artifacts. In this chapter we discuss noise issues mainly in the context of optical tweezers experiments, but much of the discussion applies to other micromechanical experiments as well.

Optical tweezers, also known as laser trapping, is a micromechanical technique that is finding increasing use in a broad spectrum of experiments in biology. Optical trapping of particles uses the momentum transfer from light scattered or diffracted by an object immersed in a medium with an index of refraction different from its own (Ashkin, 1992; Ashkin *et al.,* 1986; Ashkin and Gordon, 1983). For objects much larger than the wavelength of light, for which geometric optics is a good approximation, force is imparted by refraction and reflection. For very small objects, however, the net force is proportional to the gradient of light intensity, pointing in the direction of increasing intensity. Three-dimensional trapping of particles, large or small, can be achieved at the focus of a laser beam if a strong enough gradient of intensity can be established in all directions. To achieve relatively large trapping forces, intense laser light is brought to a tight focus by a high numerical aperture (NA) lens in a microscope; for maximal force, the particles to be trapped should be roughly matched in size to the laser focus. To minimize radiation damage in biological samples, near-infrared lasers with wavelengths of approximately 1 μm are often used; these have a focus size of approximately 0.5 μm. Typical forces that can be achieved, using up to 1 W of laser power, are on the order of tens of piconewtons (pN) (Svoboda and Block, 1994a). In the simplest applications optical traps are used, literally like a pair of tweezers, to hold and move objects such as chromosomes or organelles, or to manipulate probes such as latex or glass beads. In such cases considerations of noise are largely irrelevant. In a growing number of experiments, however, laser tweezers are used in a quantitative way both to exert or measure small forces and to measure small displacements of moving objects, with sufficient resolution to study individual biological macromolecules (DNA and RNA, or proteins). Ordinary light microscopy, limited by the wavelength of light, usually cannot provide the nanometer-scale resolution needed to observe the activity of individual molecules. While spectroscopic and scattering methods do provide molecular information from a large ensemble, they cannot easily examine single-molecule motions.

Besides optical tweezers only a few recently developed techniques such as atomic force microscopy (AFM) (Radmacher *et al.,* 1995; Rugar and Hansma, 1990; Thomson *et al.,* 1996), single-molecule fluorescence microscopy (Funatsu *et al.,* 1995; Sase *et al.,* 1995) or near-field optical microscopy (NSOM) (Betzig and Chichester, 1993) can be used to observe the dynamics of single molecules in aqueous conditions and at room temperature. Nonimaging detection, typically with fast photodiodes, can use intense illumination in many ways to track the motion of objects with Å accuracy (Bobroff, 1986; Denk and Webb, 1990). This, for example, is how the motion of an AFM cantilever is detected (Rugar and Hansma, 1990). In the case of optical tweezers, the trapping laser beam itself can be used for position detection (Svoboda and Block, 1994a). Furthermore, the trapping forces that can be exerted are on a useful scale for single-molecule experiments, for example, to stall motor proteins (Svoboda and Block, 1994b; Svoboda *et al.,* 1993) or to stretch DNA (Smith *et al.,* 1996; Yin *et al.,* 1995).

In single-molecule optical tweezers experiments, just as with any other highly sensitive method, fighting noise in its various forms becomes of foremost importance. Noise appears in electronic components, but is also unavoidably present as the Brownian motion of the observed objects, which are typically immersed in room-temperature aqueous solutions. On the one hand unavoidable noise sources set fundamental limits to micromechanical measurements. On the other hand, one can also exploit Brownian motion to calibrate the measuring apparatus itself.

This tutorial includes the following parts: Section II, a basic discussion of power spectral analysis; Section III, a derivation of the spectral characteristics of Brownian motion of optically trapped particles and a practical recipe for the way this motion can be used to calibrate optical tweezers; Section IV, a discussion of the fundamental limits of what can be measured by optical traps or other micromechanical devices; and Section V, a discussion of instrumental design techniques that will maximize the signal-to-noise ratio.

II. Spectral Data Analysis

In the type of experiments discussed here measurements are usually taken as a set of time-domain data, for example as a series of voltage measurements corresponding to the varying light intensity detected with a photodiode (Fig. 1). Time-domain data are clearly necessary to detect singular events, but a frequency-domain description of the same data has substantial advantages for interpreting "continuous" phenomena, such as oscillations and random noise signals. Experimental or thermal noise is best characterized by its power spectrum, which is a specific frequency-domain description of an original time-domain signal. Later discussion in this chapter shows how to calculate the power spectrum numerically. Further details on the calculation of the power spectrum can be found in the literature (Press, 1992).

Conceptually, a power spectrum is obtained by passing a signal (such as a fluctuating voltage from a photodiode detector) through a set of narrow-band filters and plotting the measured intensities as a function of the filters' center

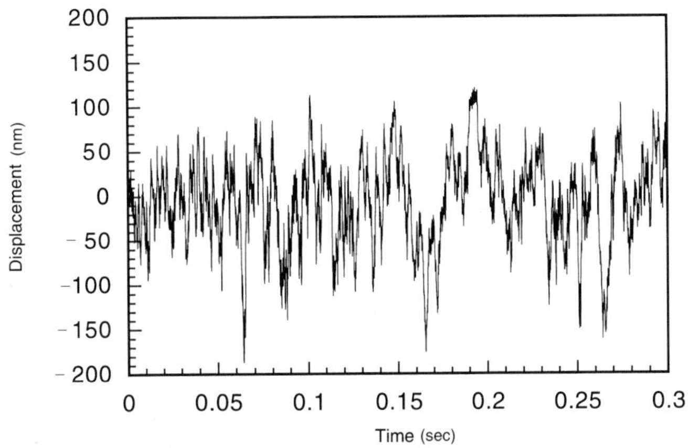

Fig. 1 Time series of data showing the Brownian motion in water of a 0.5-μm silica bead within an optical trap, at a laser power of about 6 mW in the specimen. Bead displacement was detected using an interferometric technique (Svoboda *et al.*, 1993) with a bandwidth of 50 kHz. Displacement calibration was obtained from the Lorentzian power spectrum (Fig. 4) using the methods described in this chapter.

frequencies. This process, as contrasted with a simple Fourier transform, does not preserve the total information content of the original data, as explained later. To characterize an experiment, it is necessary to know the spectral characteristics of noise, the signal, and the detection system.

A. Interpretation of the Power Spectrum

In general, going back and forth between time- and frequency-domain representations is accomplished by performing Fourier transforms. The Fourier transform of a set of real numbers (time-domain data points) gives a set of complex numbers, preserving all the information inherent in the original data. Often, however, it is more convenient to sacrifice some information content (the phases) and calculate the power spectrum or power spectral density (PSD), denoted here by $S(f)$. For practical purposes, $S(f)$ is obtained by taking the squared magnitude of the Fourier transform. This function, however, is extremely erratic: The standard deviation of each point is typically equal to its mean value. To obtain a smoother curve, many data sets must also be averaged (Press, 1992). It is this smoother curve, in the limit of infinitely many data sets, that shows the true spectral characteristics of the observed process. To understand the statistical meaning of the power spectrum, consider a set of data points, x_n: The total spread in this set of numbers is given by its variance, Var(x). One way of looking at the power spectrum is as a breakdown of this signal variance in components at frequencies f. The function $S(f)$ assigns a "power" to every frequency f, and all of the powers for nonzero frequencies add up to give exactly Var(x).

In practice, two important concepts are needed to correctly interpret $S(f)$ as calculated from a data set. These are aliasing and windowing (Press, 1992).

1. Sampling, Aliasing, and the Nyquist Frequency

If data is taken, ideally, as a series of instantaneous samples at a frequency f_s, the highest frequency component that can be unambiguously measured in the data is equal to $f_{Nyq} = f_s/2$. This f_{Nyq} is called the Nyquist frequency. A wave with frequency f_{Nyq} can have exactly one data point taken on its crests and one in its troughs. As illustrated in Fig. 2A, any wave of a frequency higher than f_{Nyq} can be erroneously interpreted as having a frequency lower than f_{Nyq}. In the power spectrum $S(f_m)$, power spectral density at frequencies above f_{Nyq} will be folded back to lower frequencies f_m below the Nyquist frequency, as shown in Fig. 2B. Such folding back of power into low frequencies is called aliasing, and the way to avoid it is to low-pass filter the signal *before* sampling it, with a cutoff frequency just at the Nyquist frequency (Horowitz and Hill, 1989).

2. Windowing

A Fourier transform of a set of N data points, used to compute the power spectrum, implicitly treats the data set as if it wrapped around periodically (i.e., mathematically x_N is implicitly followed by x_1). This can create a problem. If the data consist of, say, a pure sine wave, a narrow peak ideally is expected to appear in $S(f)$ at the wave's frequency. But the implicit wrapping around in the calculation causes the wave to appear discontinuous unless the time window is an integer multiple of the period (Fig. 3A). This discontinuity causes side lobes on the peak, as shown in Fig. 3A, which can obscure features in the power spectrum, especially close to strong lines. No perfect cure for this is possible,

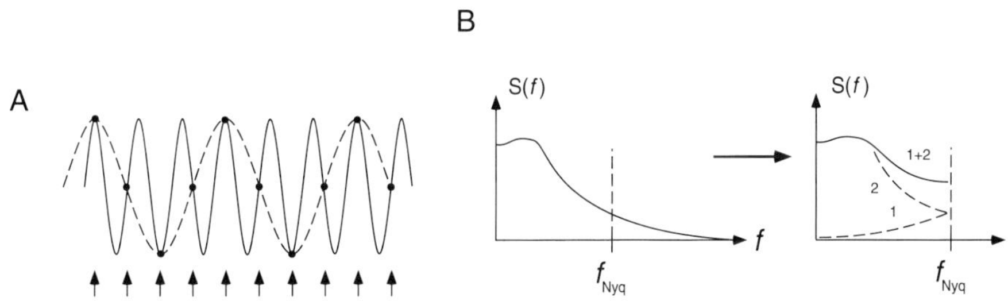

Fig. 2 (A) Schematic illustration of aliasing. A sinusoidal signal (solid curve) has a frequency that is $\frac{3}{4}$ of the sampling frequency f_s (arrows). This will falsely contribute to the power spectrum at a frequency $f_s/4$ because the sampled data (solid circles) appear as if they were produced by a wave of frequency $f_s/4$ (dotted curve). (B) For a continuous spectrum, the part of the power spectrum that continues past the Nyquist frequency $f_{Nyq} = 0.5\ f_s$ is folded back (curve 1) and added to frequencies below it (curve 2) to produce the aliased spectrum (curve 1 + 2).

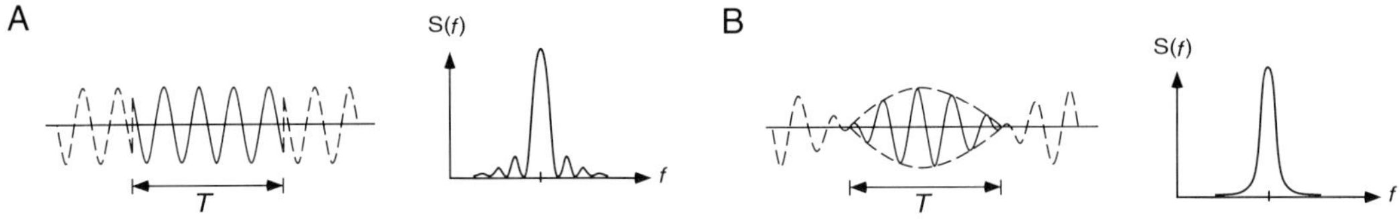

Fig. 3 Windowing of data. (A) The bottom part of the graph shows a sinusoidal signal (solid) that is measured over a time T. Outside this interval, the sinusoidal signal may continue forever. The Fourier transform algorithm, however, applied to the finite interval implicitly treats this data set as if it repeated itself with a period T (dotted curves). The artificial discontinuities and phase shifts introduced by this periodic continuation determine the width of the peak in the power spectrum (top) and create oscillations. (B) Windowing the data, that is, multiplying the data by an envelope that approaches zero at the ends of the interval (bottom), removes the oscillations in the power spectrum (top). The width cannot be reduced much.

but to minimize the effect, a "window" is applied to the data before transforming it (Fig. 3B): This means that each x_n in the data set is multiplied by some function $B(n)$ that goes to 0 at the ends of the data set. $B(n)$ should also be normalized so that the sum of all the $B(n)^2$ is equal to 1. In this way the variance of windowed data *on average* will be equal to the variance of unwindowed data (although for any particular data set, windowing changes the variance). The window shape can be optimized for specific situations but is not terribly important for relatively smooth spectra such as those discussed later. Possibilities include a parabolic hump (Welch window) or a simple triangle (Bartlett window) (Horowitz and Hill, 1989).

B. Calculation of the Power Spectrum

The following discussion shows how to obtain the power spectrum via a Fourier transform. From a set of N discrete data points x_n separated by δt, we obtain N independent fourier components $X(f_m)$, which are complex numbers given by

$$X(f_m) = \sum_{n=1}^{N} x_n e^{2\pi i n m/N}, \qquad (1)$$

where each resulting $X(f_m)$ corresponds to the frequency

$$f_m = m\delta f, \quad -N/2 \le m \le N/2. \qquad (2)$$

Before calculating $X(f_m)$ in Eq. (1), one already would have multiplied the x_n by a windowing function as described earlier. The Fourier transform in Eq. (1) is, for large data sets, greatly accelerated by use of the fast fourier transform (FFT) algorithm (Press, 1992). However, to use this algorithm, the number of data points must be an integer power of 2, a fact that should be taken into account when collecting data. Software written to calculate power spectra may revert to much slower algorithms when the data set is not a power of 2.

The frequency resolution is determined by the total length of the measurement:

$$\delta f = \frac{1}{N\delta t}. \tag{3}$$

If, as commonly is the case, the x_n are real numbers, the components $X(f_m)$ and $X(-f_m)$ are complex conjugates, and have the same modulus. The power spectrum $S(f_m)$ is calculated from the squares of these moduli. To work with positive frequencies only, the so-called one-sided power spectrum is calculated as follows:

$$\begin{aligned} S(f_m) &= \frac{2}{N^2\delta f}|X(f_m)|^2, \\ &\quad 0 < m < N/2, \\ S(f_0 = 0) &= \frac{1}{N^2\delta f}|X(0)|^2, \\ S(f_{N/2}) &= \frac{1}{N^2\delta f}|X(f_{N/2})|^2. \end{aligned} \tag{4}$$

The highest frequency in the PSD is $f_{N/2}$, the Nyquist frequency. The power spectrum consists of $N/2$ independent numbers running from $S(0)$ to $S(f_{N/2})$, even though there were N original data points: Half the original information (concerning phases) is therefore lost in the process of calculating the PSD.

From Eqs. (1) and (4) it follows that $S(0)\delta f$ is equal to the square of the average of the measured signal x_n:

$$S(0)\delta f = \bar{x}^2, \tag{5}$$

and that the sum over the power spectrum is equal to the average of the squares of the signal data:

$$\sum_{m=0}^{N/2} S(f_m)\delta f = \overline{x^2}. \tag{6}$$

Thus we obtain the relationship to the variance as mentioned earlier:

$$\sum_{m=1}^{N/2} S(f_m)\delta f = \overline{x^2} - \bar{x}^2 = \mathrm{Var}(x). \tag{7}$$

From Eq. (7) the units of $S(f_m)$ can be read off: They are $[x]^2$/Hz. It is important to keep track of numerical factors (N and 2, etc.) so that $S(f_m)$ is properly normalized to fulfill Eq. (7). There is considerable variety in the literature and in software written to calculate power spectra. It is therefore a good idea to check the normalization by directly computing the variance of a data set (*after* multiplication by the windowing function) and comparing it with Eq. (7).

III. Brownian Motion of a Harmonically Bound Particle

For the types of microscopic systems discussed here (e.g., small optically trapped particles in a solution), the theory of Brownian motion is relatively simple because linear response theory can be used, which assumes that deviations from equilibrium positions are small. In this case, the fluctuation–dissipation theorem (Landau *et al.,* 1980; Reif, 1965) states that thermal fluctuations, such as diffusion, are governed by the same parameters that apply to larger scale motions, such as sedimentation. Furthermore, in systems with a low Reynolds number (e.g., small particles moving not too fast, in a viscous medium) viscous drag is dominant over inertial forces (Happel and Brenner, 1983).

A. Power Spectrum of Brownian Motion

A particle that can move freely in a viscous fluid performs a random walk (Brownian motion) due to the continuous bombardment by the solvent molecules (i.e., it diffuses through the fluid). In accordance with the fluctuation–dissipation theorem, the diffusional motion can be predicted once the hydrodynamic drag coefficient, γ, for steady motion is measured. This is the Einstein expression for the free diffusion coefficient D (Reif, 1965):

$$D = \frac{k_B T}{\gamma}. \tag{8}$$

In terms of D, each coordinate $x(t)$ of a diffusing particle is described by

$$Var[x] = \overline{x^2} - \bar{x}^2 = 2Dt. \tag{9}$$

For three-dimensional diffusion, squared distance from the origin grows as $r(t)^2 = 6\ Dt$ because $r^2 = x^2 + y^2 + z^2$. The random excursions of the particle from its starting point grow larger and larger as time goes by. Such random diffusion, according to Eq. (8), is proportional to the absolute temperature T.

In contrast, a particle in an optical trap feels not only random forces from solvent molecules, but also a restoring force confining it within the trap and preventing long-range diffusion. As a compromise the particle will wiggle in the trap with an average amplitude that depends on the trap strength and the temperature. Near the stationary point of the laser tweezers, the trapping force will be proportional to displacement, as for a harmonic spring. Taking, for example, a 0.5-μm silica bead, the effective spring constant can be increased from 0 to approximately 1 pN/nm by varying the laser power to a maximum of approximately 1W. The position of the particle within the trap can be monitored with Å accuracy using photodiode detection (Svoboda *et al.,* 1993). At these scales of force and distance, random Brownian motion is easily visible (Fig. 1). Thermal fluctuations are characterized by an energy on the order of $k_B T$ (k_B is Boltzmann's constant), a fact that can be used to estimate the size of Brownian motions. For the case of a harmonic potential or linear restoring force, the prediction is precise: A particle trapped with a spring constant κ will have its position $x(t)$ vary according to a Gaussian distribution, with a displacement variance (Reif, 1965):

$$\mathrm{Var}(x) = \overline{x^2} - \overline{x}^2 = \frac{k_B T}{\kappa}. \tag{10}$$

At biological temperatures, $k_B T$ is approximately 4×10^{-21}N nm. In a trap with a stiffness of 1×10^{-2} pN/nm, according to Eq. (10), a particle moves randomly with an root mean square amplitude of approximately 20 nm. Therefore, the well-defined characteristics of Brownian motion can be exploited to calibrate the viscoelastic parameters of microscopic measurement devices (e.g., the spring constant of optical tweezers).

The power spectrum of the motion of a particle in an optical trap can also be calculated, which turns out to have a Lorentzian shape (Wax, 1954). An approximate equation of motion for the position $x(t)$ of the trapped particle is a Langevin equation. With a random thermal force $F(t)$ (see Reif, 1965 for a general discussion of Langevin equations),

$$\gamma \frac{dx}{dt} + \kappa x = F(t). \tag{11}$$

Equation (11) states a balance of forces, in which a drag force (friction times velocity) and a spring force (spring constant times displacement) are balanced by the random force $F(t)$ from the solvent bombardment. This is an approximation, with subtleties hidden in the random force and the friction coefficient (Wax, 1954), but in practice it describes the Brownian motion of micrometer-size objects in water very well. The random force $F(t)$ has an average value of 0, and its power spectrum $S_F(f)$ is a constant (i.e., it is an ideal white noise force):

$$\overline{F(t)} = 0 \text{ and } S_F(f) = |F(f)|^2 = 4\gamma k_B T. \tag{12}$$

Here $F(f)$ denotes the Fourier transform of $F(t)$. In writing $S_F(f) = |F(f)|^2$, and throughout the following derivation, we do not explicitly show the averaging needed to obtain $S_F(f)$ without encountering infinite integrals. From the Langevin Eq. (11), the power spectrum of the displacement fluctuations $S_x(f)$ of a trapped object can be derived. If the Fourier transform of $x(t)$ is $X(f)$:

$$x(t) = \int_{-\infty}^{\infty} X(f) e^{-2\pi f i t} df, \tag{13}$$

then the transform of $dx(t)/dt$ is $-2\pi i f X(f)$. The Fourier transform of both sides of the Langevin Eq. (11) gives accordingly,

$$2\pi\gamma\,(f_c - if)\,X(f) = F(f), \tag{14}$$

where we define $f_c = \kappa/2\pi\gamma$; f_c is the characteristic frequency of the trap. Both sides of Eq. (14) are complex expressions. By taking their squared modulus and writing $S_x(f) = |X(f)|^2$ and $S_F(f) = |F(f)|^2$ it will be found that

$$4\pi^2\gamma^2\,(f_c^2 + f^2)S_x(f) = S_F(f). \tag{15}$$

Inserting Eq. (12), the power spectrum of $x(t)$ is

$$S_x(f) = \frac{k_B T}{\gamma\pi^2\,(f_c^2 + f^2)}. \tag{16}$$

Equation (16) shows that a Lorentzian function describes how fluctuations are distributed over different frequencies f. The characteristic frequency (or corner frequency) f_c divides the Brownian motion into two regimes. For frequencies $f << f_c$, the power spectrum is approximately constant, $S_x\,(f) \approx S_0 = 4\gamma k_B T/\kappa^2$, which reflects the confinement of the particle. At higher frequencies, $f >> f_c$, $S_x(f)$ falls off like $1/f^2$, which is characteristic of free diffusion. Over short times the particle does not "feel" the confinement of the trap.

B. Trap Calibration from a Power Spectrum

Both the effective spring constant κ of an optical trap and the drag coefficient γ of the particle within it can be determined from a recording of Brownian motions and the calculation of their power spectrum. In practice, the time resolution of the detection device has to be better than the inverse of the corner frequency f_c. For typical trap strengths, this excludes video rate detection. For a laser power of 50 mW (at a wavelength of 1064 nm) in the specimen, and a 0.5-μm silica bead in room-temperature water, a typical spring stiffness is about 1.5×10^{-2} pN/nm, which results in a corner frequency of $f_c \approx 500$ Hz. Fig. 4 shows an experimental power spectrum in a double logarithmic plot.

The Lorentzian Eq. (16) depends on two parameters, κ and γ, which can be obtained by fitting Eq. (16) to the data by using, for example, the Levenberg-Marquardt algorithm (Press, 1992). Curve-fitting algorithms are implemented in many data graphing software packages (e.g., Origin for PC, Kaleidagraph for Mac, or XMGR for Unix). However, it is often convenient to roughly estimate these parameters by hand from a log–log plot of the Lorentzian spectrum:

1. The low-frequency portion of the log–log spectrum should be horizontal, but $S(f)$ may become large at the lowest frequencies due to drift and low-frequency vibrations. First, draw a horizontal line that ignores such effects and call its height S_0.
2. The high-frequency portion of the spectrum should be a line of slope approximately −2. Draw this line and extend it to intersect the horizontal S_0 line; this intersection determines f_c, the "corner frequency."

Once S_0 and f_c are measured, the trap stiffness can be calculated as

$$\kappa = \frac{2k_B T}{\pi S_0 f_c}, \tag{17}$$

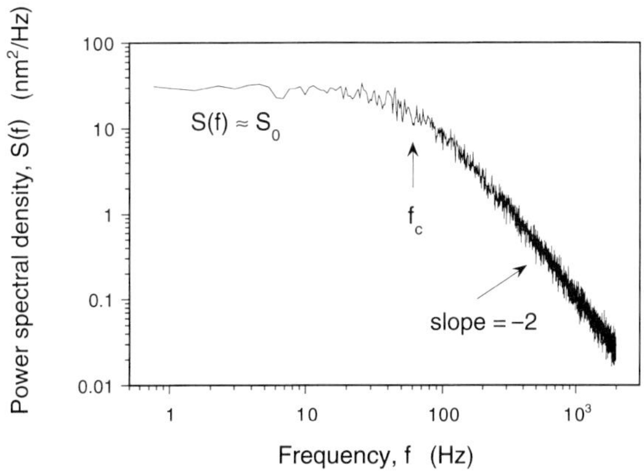

Fig. 4 The Lorentzian power spectrum of the Brownian motion of a 0.5-m silica bead moving within an optical trap at a laser power of 6 mW in the specimen, obtained from a time series (partially shown in Fig. 1) of voltage readings from an interferometric detector (bandwidth 50 kHz). About 30 spectra from independent intervals of the original time series were averaged. The corner frequency is $f_c \approx 60$ Hz and the plateau power $S_0 = 0.028\ V^2/Hz$. In this case, the theoretical drag coefficient of the sphere (Eq. 20) was used to determine the trap stiffness, $\kappa = 1.7 \times 10^{-3}$ pN/nm, and to calibrate the response of the detector (32 nm/V).

and the drag coefficient γ of the particle is

$$\gamma = \frac{k_B T}{\pi^2 S_0 f_c^2}. \tag{18}$$

If γ is known from first principles (see later), κ can be calculated directly from the corner frequency f_c:

$$\kappa = 2\ \pi \gamma f_c. \tag{19}$$

By using Eq. (10), an attempt could be made to estimate $\kappa = k_B T/\text{Var}(x)$ directly from the variance of the data set without examining the power spectrum. But this is risky because very low-frequency noise from drift, vibrations, or other sources will often artificially inflate Var(x). The advantage of plotting the power spectrum is that such effects are often easily recognizable; estimating S_0 and f_c generally gives a better value for κ. When κ has been determined from S_0 and f_c, a better estimate for Var(x) can, if needed, be calculated from Eq. (10). Similarly, instrumental noise at high frequencies may eventually cause the sloping spectrum to level off again, but this usually happens at an amplitude low enough not to affect the fit parameters.

C. Hydrodynamic Drag

It is often desirable to calculate the viscous drag coefficient γ of a particle from first principles, for example to compare with values estimated from a thermal-noise power spectrum. Using Eq. (8), the object's free Brownian motion can also be predicted when γ is known. The hydrodynamic drag coefficient also needs to be known for calibrating trapping forces by sweeping the trap through the fluid and observing the particle displacement within the trap (Svoboda and Block, 1994a).

To calculate γ theoretically, a hydrodynamic problem must be solved. This is usually difficult, even when inertia is negligible at a low Reynolds number. The most important practical case was solved long ago and has a simple result: It is the Stokes drag on a small sphere far from any surface (Reif, 1965):

$$\gamma = 6\pi\eta a. \tag{20}$$

Here η is the dynamic viscosity of the solvent and a is the radius of the sphere.

There are many exact and approximate formulas giving γ for various particles in unbounded solutions (Happel and Brenner, 1983), and these apply to trapped particles as well. A complication often arises in microscopy experiments when the observed object is close to a sample chamber surface. For a particle close to a surface—at a distance similar to or less than its diameter (a)—the unbounded-solution drag coefficients are no longer correct and cannot be used to predict Brownian motion. Drag near a surface is due largely to shear between the particle and the wall, which is a different hydrodynamic situation from shear flow around a free particle. This remains true when flow is induced above a surface (so that velocity increases in proportion to the height above the surface); an unbounded-solution γ cannot be combined with a local velocity to obtain the drag force.

For a sphere in the vicinity of a surface, but still with $a/h < 1$, the correction to first order in a/h to Eq. (20) is

$$\gamma = 6\pi\eta a\left(1 + \frac{9}{16}\frac{a}{h}\right), \tag{21}$$

which applies to horizontal motion (parallel to the wall) with the sphere center at a height h. Moving vertically, toward or away from the wall, the factor in Eq. (21) becomes $(1 + (9/8)a/h)$. Equation (21) is known as the Lorentz formula (Happel and Brenner, 1983).

IV. Noise Limitations on Micromechanical Experiments

Micromechanical experiments measure forces and displacements produced by microscopic objects. Such measurements are typically done by monitoring small deformations or displacements $x_p(t)$ of an elastically suspended probe as it interacts with the object (Fig. 5). One example of such a probe is a particle in an optical trap in which, typically, probe motion is followed as a function of time. In atomic force microscopy (AFM) experiments, surfaces are imaged by scanning a sharply pointed, elastically suspended tip laterally across the surface and then converting the time series data of tip deflection into a spatial image. In an optical

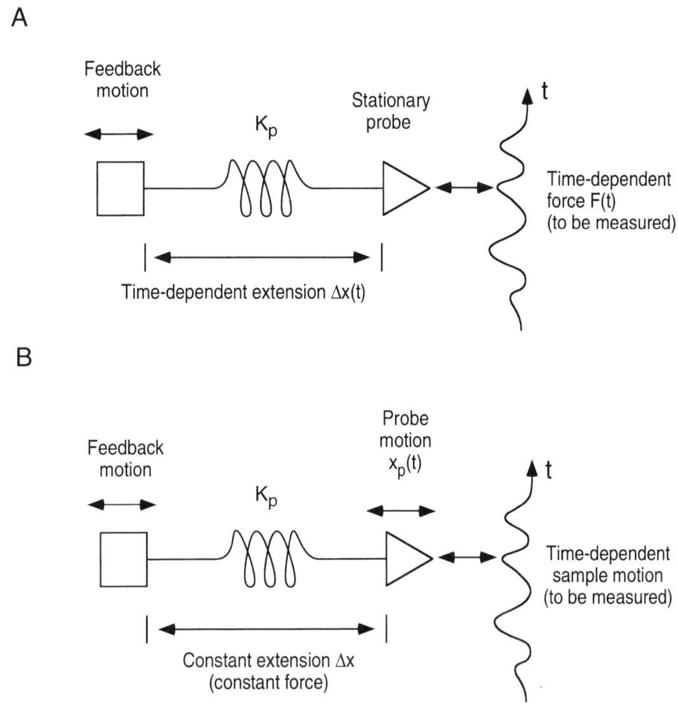

Fig. 5 Schematic representation of two prototypical micromechanical experiments. In an optical trap, the probe (triangle) is a trapped dielectric particle; the anchor point (square) represents the position of the center of the trap, which is controlled through feedback; and K_p represents the trap stiffness. In atomic force microscopy, the probe is the scanning tip; the anchor point is the base of the cantilever (controlled via feedback); and K_p represents the cantilever stiffness. The probe interacts with the sample through a force that, in general, changes with distance and time. (A) Position-clamp experiment to measure force. The absolute probe position is monitored with high precision, and the anchor point is moved to keep the probe stationary. From the changing distance between the probe and the anchor point (the probe strain), the changing force on the stationary probe can be deduced. (B) Force-clamp experiment to measure position. The probe position is again monitored, and the anchor point is moved to keep the probe strain, and thus the force on the probe, constant. The anchor motion then reflects how the sample moves under a fixed, constant force. The probe response is low-pass filtered by the dynamic response characteristics of the probe as described in Section IV, C of this chapter. If the probe is scanned along a surface (AFM), a constant-force contour, within the limitations of the probe dynamic response, is traced by the anchor motion.

trap, the displacement Δx of the particle away from the trap center $x_0(t)$ is measured (i.e., $\Delta x(t) = x_p(t) - x_0$ where $x_p(t)$ is the instantaneous position of the probe). In AFM, a tip displacement Δx represents the distortion of the elastic cantilever that supports the tip. In either case, we can call the relative displacement $\Delta x(t)$ the *probe strain.* If the stiffness K_p of the elastic element (the *probe stiffness*) has been calibrated, the suspension force on the probe is inferred as

$$F(t) = K_p \Delta x(t). \tag{22}$$

In scanned-tip AFM experiments and in some optical tweezers applications, feedback is used in conjunction with position detection; the anchor point of the "spring" holding the probe (e.g., the base of the cantilever or the center of the trap) can be moved quickly and precisely. There are then two prototypical experiments that can be performed—although actual experiments, or experiments done without feedback, may be intermediate between these two cases. In one case, force is measured (i.e., probe strain $\Delta x(t)$ is monitored) while feedback keeps the probe at an absolutely fixed position $x_p(t)$ = constant. This is a *position clamp* or isometric experiment. The other prototypical experiment measures probe motion at an absolutely constant force: Feedback keeps the probe strain Δx constant as the probe itself is moved by its interaction with the object. This is a *force clamp* or isotonic experiment. Force clamps and position clamps may seem to be unattainable idealizations, but, in fact, present technology, using piezoelectric actuators (in AFM) or acousto-optic or electro-optic modulators (with optical tweezers), can approximate ideal conditions quite well up to high frequencies. In both the position clamp and the force clamp, the anchor point of the probe assembly is moved by the feedback circuitry, but according to different criteria, keeping $x_p(t)$ constant in the first case and $\Delta x(t)$ constant in the second case. Different types of detectors are necessary for these two types of experiments. We next discuss how in these two cases the sensitivity of micromechanical measurements is limited by thermal noise.

A. Position-Clamp Experiments

First we consider pure force measurements by means of position clamping, in which one wants to measure a time-varying force signal $F_{sig}(t)$ on the probe. This would not normally be done with an AFM, but in an optical trap, for example, the force production of a molecular motor tied to a stationary load can be measured. When the varying force generated by the object begins to displace the probe, the equilibrium position $x_0(t)$ is quickly changed by moving the trap, changing the probe strain $\Delta x(t) = x_p - x_0(t)$ to balance the varying force and keep the probe at a fixed position, x_p. The total time-dependent force exerted on the probe is found from the observed $\Delta x(t)$:

$$F_{tot}(t) = K_p \Delta x(t). \tag{23}$$

Assuming that feedback control of the probe position is perfect, the fundamental limitation in measuring the force the object exerts on the probe comes from the presence of a white-noise thermal force that acts on the probe in competition with the force to be measured. From Eq. (12), the power spectrum of this thermal force is

$$S_F(f) = 4\gamma k_B T, \tag{24}$$

where γ is the frictional drag coefficient on the probe. In optical trapping, all friction comes from hydrodynamic drag on the trapped particle; in AFM, γ

includes all friction opposing the motion of the tip (i.e., drag on the tip and on the cantilever).

The practical implication of Eq. (24) is that the relative noise level can be decreased by low-pass filtering of the strain signal $\Delta x(t)$ with a cutoff frequency greater than the fastest rate of change in the signal. Because the force noise is distributed evenly over all frequencies, such filtering can increase the signal-to-noise ratio substantially. This argument assumes that the detection is fast enough to follow the force signal in the first place. How accurately can $F_{sig}(t)$ be determined at any particular time? As a concrete example, assuming fast enough detection, consider a force signal $F_{sig}(t)$ produced by a molecular motor: the slower the true signal varies, the better it can be resolved; the lower the permissible cutoff frequency of the low-pass filter, the more noise is removed. The remaining uncertainty $\Delta F(t) = F_{tot}(t) - F_{sig}(t)$ with a properly chosen filter frequency f_s corresponds, on average, to the integrated noise power below f_s, which is equal to the constant noise spectral density in Eq. (24) multiplied by the frequency range, 0 to f_s, passed by the filter,

$$\Delta F_{rms} = \sqrt{\overline{\Delta F^2}} = \sqrt{4\gamma k_B T f_s}. \tag{25}$$

Equation (25) states the fundamental resolution limit of a pure force measurement. It shows that the measurement can be optimized by either reducing the drag γ on the probe or keeping the rate of change of the true force signal f_s as low as possible (e.g., by scanning slowly with an AFM). Note that the stiffness of the elastic probe suspension is *not* relevant in principle. However, in practice, the noise in the strain detector electronics limits how small a strain in the probe can still be detected. Therefore, a softer probe allows measurement of both a smaller force change and a smaller absolute force. As a rule of thumb, the sensitivity of a detector is large enough or its noise contributions are low enough when the thermal motion of the probe can be detected. Increasing the sensitivity beyond this point brings no advantage.

Making sure that the force signal varies slowly, that is, reducing f_s in Eq. (25), may not always be possible. Nevertheless it is always advantageous to low-pass filter the signal to the lowest possible frequency. For static forces, $f_s = 0$ can, in principle, be measured to arbitrary precision with correspondingly long measurement times. In practice, however, measurements of static or very slowly varying forces are limited by drift in the apparatus, not by Brownian noise. In some cases the detection system is intrinsically slower than the variation of the signal. This typically happens when video recording and image processing is used for displacement detection. In that case it must be remembered that the detected position is a time average and may not reflect the true excursions of the probe.

B. Force-Clamp Experiments

Now we consider pure probe position measurements at constant force. The probe strain Δx is kept constant, corresponding to a constant suspension force

$F_{set} = K_p\, \Delta x$ on the probe that is balanced by the force of interaction with the sample. With an AFM it is possible, for example, to trace a surface or the shape of a biological macromolecule as defined by its constant-force contours. In an optical trap, the motion of a molecular motor under a constant load force can be followed.

We neglect, for the moment, viscous drag effects, which are treated in part C. In other words, we assume that the sample and the probe move very slowly. If thermal noise were absent, a force-clamp apparatus would allow a force value to be dialed in, and the probe would always exert *exactly* that force on the sample. The probe position $x_p(t)$ would exactly trace a constant-force contour of the object (in AFM) or follow exactly the motion of the motor protein under constant load (with optical tweezers). In reality, with thermal noise the force exerted by the elastic suspension of the probe (i.e., the optical trap or the cantilever) is balanced by the sum of sample interaction force and the fluctuating thermal force on the probe. The probe position $x_p(t)$ is then only an estimate for the true constant-load position corresponding to the dialed-in force.

There are two experimental goals that need to be distinguished at this point. Figure 6 illustrates the situation with a hypothetical interaction force profile between probe and sample. For example, this could be a plot of how the repulsive force increases when the tip of an AFM gets closer to a surface, or a plot of how the attractive force in the elastic linkage between a probe bead and a molecular motor increases with increasing distance. Thermal noise imposes dis-

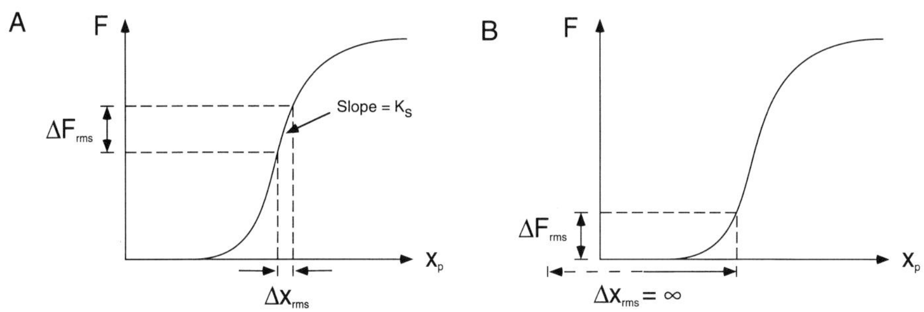

Fig. 6 Force-clamp experiments at high force and at low force (edge detection). The solid curve is an instantaneous force profile $F(x_p)$ as a function of probe position x_p. An uncertainty ΔF_{rms} in force measurement results from thermal forces on the probe. (A) The goal is to monitor changes in $F(x_p)$ with time, or equivalently to measure a spatial constant-force contour with a scanned probe (AFM). The apparatus is operating in a constant-force mode at values F_{set} well above the force uncertainty $F_{set} >> \Delta F_{rms}$. The force uncertainty translates, via the slope K_s of $F(x_p)$, into an uncertainty in locating the position $x(F_{set})$ on the force profile corresponding to the set force. The force resolution ΔF_{rms} of the apparatus is given by Eq. (27). (B) Locating the "edge" of a profile in the least invasive manner (i.e., using the smallest possible force). In any real system the interaction force will smoothly approach zero at some distance. The smallest possible set force is $F_{set} \approx \Delta F_{rms}$. If F_{set} approaches ΔF_{rms} the position uncertainty diverges to infinity.

tinct limitations in two experimental situations: (a) It limits the accuracy with which a high-force spatial response can be determined, and (b) it sets a minimum force at which a spatial response can be obtained at all. We consider each of these cases in turn.

1. Displacement Measurements with Force Clamp at Large Forces

To be specific, one might want to challenge a molecular motor with a load close to its stalling force or to image the underlying substrate of an AFM sample. The displacement response to a strong force may, of course, include some deformation of the object under study. Note that the quantity being recorded is always the probe position $x_p(t)$ for the set interaction force F_{set}. Which property of the sample this reflects varies from case to case. It could, for example, report conformational changes in a motor protein in the case of our optical tweezers example, or local differences in the surface chemistry for the AFM example. In any case, what we mean by "large" force is that F_{set} is large compared to the root-mean-square thermal force on the probe $F_{set} >> \Delta F_{rms}$ (see Eq. 25). The position uncertainty in the experiment is caused by the force uncertainty ΔF_{rms} (Fig. 6A). If the local stiffness of the probe sample interaction is K_s (i.e., the local slope of F vs. x in Fig. 6), then

$$\Delta x_{rms} = \frac{\sqrt{4\gamma k_B T f_s}}{K_s}. \tag{26}$$

This is the fundamental limit of a pure position measurement at a relatively large constant force. Again, the stiffness of the trap or cantilever does *not* enter directly, but, instead, the characteristics of the force between probe and object are determining the error. For example, in measuring the displacement $x_p(t)$ caused by the action of a molecular motor, Eq. (26) shows that the uncertainty Δx_{rms} can be very small if the stiffness K_s of the bead motor linkage is high. The thermal noise can be further reduced by decreasing the drag coefficient of the probe. It is also still true that static displacements against a finite load can be measured better the longer one takes to measure them (reducing f_s). In practice, though, mechanical drift is encountered again at long times.

2. Edge Detection

In some situations it is necessary to detect the edge of a force profile without disturbing the object. This is crucial, for example, in imaging soft biomolecules by AFM, when it is desirable to follow the lowest possible force contour. This situation is illustrated in Fig. 6B, in which the object force $F_{obj}(t)$ is shown as a curve. In any real system the interaction force between probe and sample will smoothly approach zero at some distance as shown in the figure. Again neglecting viscous drag, the suspension force on the probe F_{set} is always balanced by both the sample–probe interaction force and the random thermal force $\Delta F(t)$ given by Eq. (25): $F_{set} + \Delta F(t) + F_{obj}(t) = 0$. The thermal noise now determines the lowest force contour that can be followed. Consider, for example, how the feedback system operates for an AFM probe close to a repulsive surface. Assume that a thermal $\Delta F(t)$ pushes against the probe in the same direction as the surface. The feedback will move the probe farther away from the surface, decreasing $F_{obj}(t)$ to compensate. Now, if F_{set} is so low that $F_{set} + \Delta F(t)$ can become negative, the feedback will try to move the probe infinitely far away from the surface (i.e., it cannot compensate). In practice, this means that to locate the "edge" of a force profile, at least a force $F_{set} \approx \Delta F_{rms} = \sqrt{4\gamma k_B T f_s}$ must be applied.

For noise reduction, as in a position-clamp experiment, the primary goals are to reduce the drag on the probe and the filter frequency f_s. Surprisingly, probe stiffness again is not a direct consideration in avoiding sample deformation. In practice, however, detector resolution in the feedback circuit may become limiting, in which case a lower force clamp is possible with a less stiff probe such as a softer cantilever.

C. Dynamic Response of the Probe Interacting with a Sample

There is an important dynamic limitation for force-clamp experiments that we neglected so far and which is closely related to the preceding noise discussion. Even if feedback is perfect and the suspension force is held constant, the probe cannot respond instantaneously to the motion of the sample because of the viscous drag γ on the probe. For example, if an AFM scan is made at too high a scan rate, the probe will not follow a compliant surface, but will simply plow through a nearly constant height, yielding little information. Alternatively, if a motor protein performs a fast conformational change, the bead that holds the motor cannot instantaneously follow the change. Consequently, a force clamp cannot, in principle, follow motion perfectly if the probe has any drag at all. The probe motion is a low-pass filtered version of the object motion, and the force on the object deviates from the set value.

Assume that by using a probe with no drag ($\gamma = 0$), an "ideal" constant-force probe motion could be measured; call this $x_{p0}(t)$. In an actual measurement with probe motion $x_p(t)$, however, the drag force on the probe is $-\gamma dx_p(t)/\mathrm{d}t$, which causes the actual motion to be different from $x_{p0}(t)$. If the local stiffness of the probe–object interaction is K_s, drag force is balanced by an additional sample deformation force, which is $K_s\,(x_p(t) - x_{p0}(t))$. This balance can be written as

$$\gamma \frac{d}{dt} x_p(t) + K_s x_p(t) = K_s x_{p0}(t). \tag{27}$$

In a procedure similar to that following Eq. (11), it is found from Eq. (27) that the power spectrum $S_x(f)$ of $x_p(t)$ is related to the power spectrum $S_{x0}(f)$ of the ideal signal $x_{p0}(t)$ by

$$S_x(f) = \frac{f_{ps}^2}{(f_{ps}^2 + f^2)} S_{x0}(f). \tag{28}$$

The signal is cut off above a characteristic probe-sample frequency $f_{ps} = K_s/2\pi\gamma$. This means that signal frequencies higher than f_{ps} will be suppressed by the probe response and will be unmeasurable in practice.

For example, in experiments measuring motor protein forces, the stiffness K_s of the motor-bead linkage may be variable, between 0.01 and 0.1 pN/nm (Coppin *et al.,* 1996; Kuo *et al.,* 1995; Meyhofer and Howard, 1995; Svoboda and Block, 1994b; Svoboda *et al.,* 1993), which for 0.5-μm beads implies a cutoff frequency on the order of 1 kHz. In contrast, AFM probes against protein surfaces, which typically have elastic moduli of several GPa (Gittes *et al.,* 1993), show effective spring constants K_s on the order of 10^2pN/nm, which for a low-drag probe could lead to very high cutoff frequencies.

V. Sources of Instrumental Noise

Optical trapping experiments are most often combined with some form of light microscopy, so that the laser and the special optics required are added to a commercial microscope or integrated into a custom-built microscope (Kuo and Sheetz, 1993; Molloy *et al.,* 1995; Simmons *et al.,* 1996; Smith *et al.,* 1996; Svoboda and Block, 1994a; Svoboda *et al.,* 1993). Nanometer-scale position detection, usually of a trapped latex or silica bead, is commonly the primary measurement. From the displacement, appropriate calibration of the trapping force provides a measure for the force exerted on the bead. Several sources of instrumental noise, depending on the specific detection method, will affect the primary displacement measurement and limit both its spatial and temporal resolution.

It is often easiest to use an existing standard video system to determine bead position from its video image via fluorescence or a contrast-enhancing transmitted-light imaging method such as phase contrast or differential interference contrast microscopy (DIC). In this case, temporal resolution is limited to the video half-frame rate of 60 or 50 Hz, depending on the video system used. This usually is not sufficient to resolve dynamic processes on the level of single molecules. Spatial resolution is limited, in a complicated way, by the optics, the camera, the video storage device, the image processing method, and so forth (Gelles *et al.,* 1988; Inoue, 1986; Schnapp *et al.,* 1988). In practice it is very hard to achieve a position resolution as low as 10–20 nm for an object such as a 0.5-μm silica bead. For these reasons, most quantitative experiments are performed using nonimaging detection systems based on photodiodes. We discuss these systems in more detail.

A. Noise from Electronics

The amplifier design to be used with photodiodes depends on the conditions of the experiment. Good introductions can be found in the literature (Horowitz and Hill, 1989; Sigworth, 1995) and in photodiode manufacturer's catalogs [e.g., UDT Sensors, Inc. (Hawthorne, Ca), Advanced Photonics, Inc. (Camarillo, Ca), Hamamatsu (Hamamatsu City, Japan)]. If speed and linearity are important, and if the light levels are not extremely low, photoconductive operation with a reverse bias is best (Figure 7A). A low-noise operational amplifier acts as a current-to-voltage converter, so that the photodiode is operated as a pure current source. We now discuss sources of noise for this case. The responsivity of silicon photodiodes varies between 0.2 and 1.0 A/W for light with wavelengths between 350 and 1100 nm, with a maximum at about 1000 nm. Light levels as low as pW can be detected, at the cost of poor time resolution (i.e., low bandwidth; see later). In practice, intensities more than approximately 100 μW in typical optical-tweezers experiments result in a signal-to-noise ratio of better than 1×10^5 at a bandwidth of 100 kHz. At the high end, the maximum intensity that can be

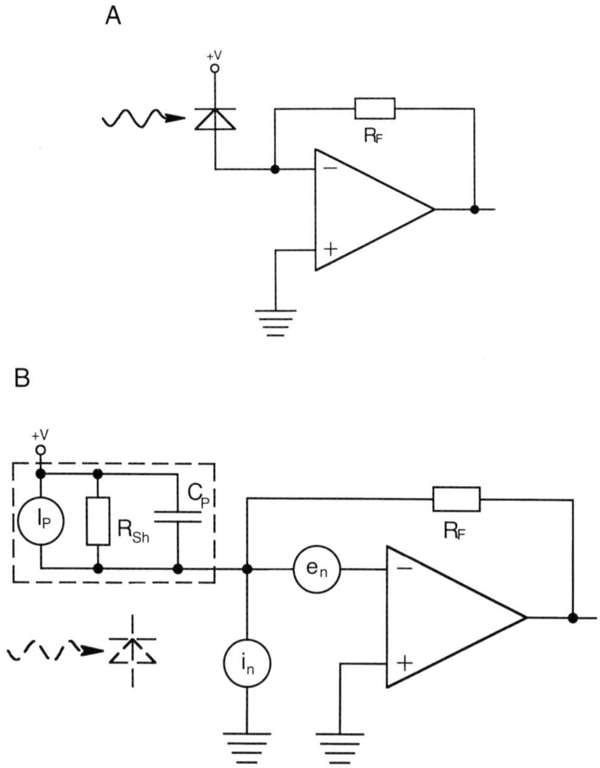

Fig. 7 (A) A typical circuit diagram for operating a photodiode in photoconductive mode with a reverse bias (current-to-voltage converter). (B) Equivalent circuit for the purpose of noise discussion. Noise sources that in reality are internal to the operational amplifier are represented by an equivalent voltage source e_n and a current source i_n acting at the inputs to the op-amp. The photodiode is replaced by an equivalent circuit including the junction capacitance, the shunt resistance, and an ideal current source.

measured depends on both the area of the detector and the width of the light beam. Deviations from linearity, to maximum intensities of approximately 1 mW/cm^2, are typically below 5%, but by 10 mW/cm^2 the response is strongly nonlinear. Linearity is improved at high light levels by using a large reverse bias in photoconductive operation.

Most of the noise sources discussed here are, at least approximately, white noise (i.e., the noise power spectrum of each is approximately constant for all frequencies). If the bandwidth (i.e., the Nyquist frequency) is given, each of these noise contributions can be calculated as a mean-squared noise current and added together—assuming they are statistically independent—to give the total mean-squared noise current. The square root of this quantity can then be compared to the photocurrent to obtain a relative noise contribution. Figure 7B shows an equivalent model circuit highlighting the noise sources discussed in the following.

1. Shot Noise

Photons are absorbed in the diode, creating electron-hole pairs and, eventually, a flow of current in the external circuit. Measuring this current amounts to counting elementary charges, which like other random counting processes (Poisson processes) results in a statistical variance equal to the number of counts, $\overline{\Delta n^2} = \bar{n}$. This gives rise to a counting noise, known as shot noise, as follows (Horowitz and Hill, 1989). Suppose there is an average photocurrent I_p, which we want to measure with a certain time resolution Δt (i.e., we count the number of electrons arriving within each sampling window Δt). As discussed earlier, a sampling time Δt corresponds to a bandwidth (Nyquist frequency) of $B = 1/(2\,\Delta t)$. The number of electrons counted in each bin, and therefore the variance in the number of counts, is $\overline{\Delta n_e^2} = \bar{n}_e = \Delta t I_p/q_e$, where q_e is the elementary electronic charge of 1.6×10^{-19} C, and I_p is the photocurrent. Therefore the variance in the current (Fig. 8) is

$$\overline{\Delta I_S^2} = \frac{q_e^2 \overline{\Delta n_e^2}}{\Delta t^2} = 2 q_e I_p B. \tag{29}$$

This result shows that shot noise is a white noise: It has a constant spectral density of $2q_eI_p$. The minimal current I_p entering this equation at the lowest light levels is the dark current of the photodiode, which for low-noise diodes is approximately 50 nA (for a 100-mm^2 diode at 10 V bias voltage). Shot noise is usually the dominating noise source if the electronics are designed carefully, using low-noise components. As an example, assume a photocurrent of I_p = 5 mA, corresponding to a light intensity of 12.5 mW at a responsivity of 0.4 mA/mW. With a bandwidth of B = 100 kHz, the root-mean-square shot-noise current from Eq. (29) is I_s = 12.6 nA, a relative contribution of 2.5 ppm in the 5 mA photocurrent.

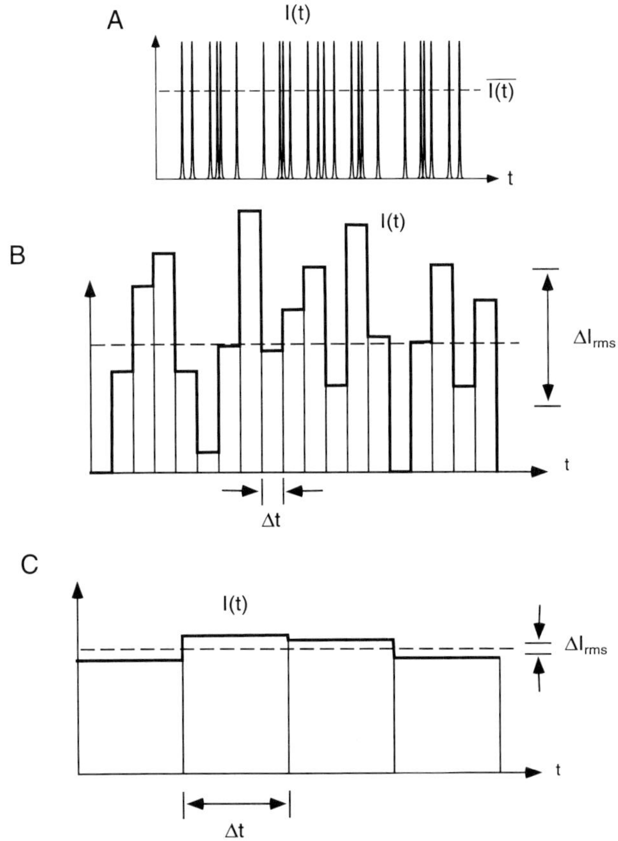

Fig. 8 The origin of shot noise. (A) A current I(t) due to the independent passage of elementary charge carriers through a point in the circuit (such as through the photodiode p-n junction) consists of a series of very narrow spikes. (B) Measuring current with a small window size Δt, and thus a large Nyquist frequency $f_{\mathrm{Nyq}} = 1/(2\Delta t)$, means that a small number of spikes are counted in each sampling time. The measured current thus has a large variance, which is the shot noise superposed on the DC current value. (C) Measuring current with a larger window size, and thus a smaller Nyquist frequency, means that proportionally more charges are counted in each sampling time. The random noise superposed on the DC current value is smaller. As shown in this chapter, the variance of this random noise is proportional to the Nyquist frequency.

2. Johnson Noise

Any resistor produces noise, called Johnson noise, through the thermal motion of its electrons. This noise appears as a fluctuating voltage across the terminals of the resistor, or as a fluctuating current if the terminals are connected by other circuitry (Horowitz and Hill, 1989). This random voltage is exactly analogous to the random force in Brownian motion, and the voltage power spectrum S_v is given by Eq. (16) except that the drag coefficient is replaced by the resistance R: $S_v = 4k_BTR$. Because S_v is a constant, Johnson noise is white noise. In our circuit (Fig. 7B), the shunt resistance R_{shunt} and the feedback resistor R_{Fbk} each contribute a mean-squared current that is their mean-squared Johnson voltage divided by the respective R^2. Multiplying by bandwidth B, gives the mean-squared Johnson current:

$$\overline{\Delta I_J^2} = \frac{4k_BTB}{R}. \tag{30}$$

Assuming $R_{shunt} = 5$ MΩ and $R_{Fbk} = 1$ kΩ, and a bandwidth of 100 kHz will give root-mean-square noise currents of $\Delta I_J(R_{shunt}) = 18$pA, and $\Delta I_J(R_{Fbk}) =$ 1.3 nA respectively, both smaller than the shot noise. Also, depending on material and construction, resistors produce some excess noise in addition to their Johnson noise. It is important to choose low-noise resistors, typically metal-film resistors, for at least the input stages of the amplifier.

3. Amplifier Noise

Operational amplifiers produce their own noise because of the shot noise and resistor noise that originate from their internal elements. It is common (e.g., in data sheets for op-amps) to express these noises as input equivalent voltage and current, that is, voltage and current at the input of an ideal op-amp that would produce the same noise at the output (Horowitz and Hill, 1989), as shown in Figure 7B. Data sheets for op-amps usually state an "equivalent root-mean-square input noise voltage" e_n and an "equivalent root-mean-square input noise current" i_n, which actually are the square roots of the power spectral densities of noise voltage and current and must be multiplied by the bandwidth B to obtain the actual root-mean-square quantities. Because in our application we want to compare all noise contributions to the photocurrent, we need to convert the amplifier noise voltage into a current using the feedback resistor R_{Fbk} and the photodiode capacitance C_P and add it to the input current noise i_n^2 to find the total amplifier noise:

$$\overline{\Delta I_{amp}^2} = \left[\frac{1}{R_{Fbk}^2} + (2\pi f)^2 C_P^2\right] e_n^2 B + i_n^2 B. \tag{31}$$

Taking data from a typical appropriate operational amplifier, AD743 (Analog Devices, Norwood, MA), $e_n \approx 3$ nV/Hz$^{1/2}$, $i_n = 6.9$ fA/Hz$^{1/2}$, ignoring the slight frequency dependence of the noise voltage, assuming $R_{Fbk} = 1$ kΩ, $C_p =$ 300 pF, and again assuming a bandwidth of 100 kHz, we calculate $\Delta I_{amp} =$ 1.0 nA. This an upper limit because we just used the smallest reactance of the diode (at 100 kHz) for the whole frequency range. Compared to the shot noise (12.6 nA, in our example) op-amp noise currents are negligible, although this may not be the case if low-quality components are used, or if the band width must be higher. The next stages in amplification usually contribute less noise than the input stage.

The relative error can, of course, change dramatically when the difference between two photodiode signals is computed. This is usually the case in position detection using, for example, a quadrant diode. It is then more meaningful to express the noise level as a minimal measurable displacement. In the displacement detection system in our laboratory, we end up with electronic noise corresponding to about 0.5Å in displacement of a 0.5-μm silica bead at 100 kHz bandwidth.

As a final stage of the electronic detection, if the measurements eventually are read into a computer, digitization errors need to be considered. The resolution of an analog-to-digital converter (ADC) is given in bits: A 16-bit ADC translates the maximal analog voltage for which it is designed (typically 10 V) into the integer number $2^{16} = 65{,}536$. Besides simple rounding error, imperfections in the circuitry usually cause the least significant bit to fluctuate between 1 and 0. If this error is independent for each sampled voltage point, the result will be white noise with root-mean-square variation of $1/65{,}536 \approx 15$ ppm, but spread out to the Nyquist frequency. Bandwidth reduction in general will decrease this error. However, ADC noise is complicated and sometimes is not even limited to the last bit, depending on the type of converter and the computer environment. In case of doubt it is best to measure the converter noise directly. ADCs with relatively few bits or an input signal not using the full dynamic range of the ADC obviously present problems.

B. Other Noise Considerations

With enough light intensity, as described earlier, photodiode detection can be used to monitor the motion of μm-size beads, with Å resolution, at bandwidths up to 100 kHz. A laser is commonly used to achieve sufficient intensity focused on a bead. In using optical tweezers, this laser can be the trapping laser itself or a separate laser. The advantage of using the trapping laser is that the detection system is intrinsically aligned with the trap and a relative displacement is measured. If absolute position needs to be measured while the trap is moved, a separate laser is needed. By using a laser focused on the trapped object, a number of new noise problems are created. Lasers show fluctuations in laser power, beam pointing, and frequency. Intensity fluctuations are usually a few percent of the maximal power and are not critical as long as the laser is operated at a relatively high power and intensity regulation for trapping or detection is performed farther down the line (e.g., with polarization optics). Most lasers are also extremely sensitive to backreflections, which can cause large-amplitude intensity oscillations. The most efficient but costly way to avoid these is to use a Faraday-effect type of optical isolator (Optics for Research, Caldwell NJ; Conoptics Inc., Danbury CT; Electro-Optics Technology, Inc., Traverse City MI). Alternative low-cost approaches are (a) placing the first reflecting surface at a distance from the laser that is larger than the coherence length of the laser; (b) Using a neutral density filter (tilted to the beam) to attenuate the transmitted and the back-reflected beam, which is only practical if there is laser power to spare; and (c) using the combination of a linear polarizer and quarter-wave plate, which produces circularly polarized light if the polarizer is oriented at 45° to the fast axis of the quarter-wave plate. Light back-reflected off a mirror has the sense of its circular polarization inverted and is blocked by the polarizer after being converted back to linearly polarized light by the quarter-wave plate. In practice this method is limited in its effectiveness because back-reflections can have phase changes other than what a plane mirror produces at normal incidence.

Beam-pointing fluctuations are a more serious problem. They are caused mainly by changing thermal gradients inside the laser. Different types of lasers show different amounts of these fluctuations. Large-frame noble gas lasers usually show fewer fluctuations than solid state lasers (Siders *et al.,* 1994). For diode lasers, data were not available from manufacturers. Among the solid state lasers, which are most often used for optical trapping, the crystalline substrates vary in thermal conductivity. Thermal lensing, caused by thermal gradients in the laser rod, is in some designs used intentionally for gain increase. Neodymium : yttrium lithium fluoride (Nd : YLF) has a large thermal conductivity and therefore less beam-pointing instabilities than neodymium : yttrium aluminum garnet (Nd : YAG). It is best to request detailed information from design engineers at the manufacturer. Beam-pointing instabilities for solid state lasers are typically up to 50 μrad, and the beam usually does not pivot around a fixed point. For trapping, the beam usually is expanded by a factor of about 5, which decreases angular fluctuations by the same factor. Assuming a typical focal length of 1.5 mm in a high-magnification microscope objective, a laser with pointing fluctuations of 10 μrad in the back focal plane of the objective will cause lateral fluctuations of the trap by about 15 nm. Using this laser for position detection would thus severely limit the resolution. The pointing fluctuations are typically quite slow, on the order of 1 Hz and slower, so that fast displacements still can be detected with better resolution. Single-mode polarization-preserving optical fibers have been used to stabilize the beam (Denk and Webb, 1990; Svoboda *et al.,* 1993). The reduction in beam-pointing fluctuations can be on the order of 10-fold, but fibers introduce their own noise problems, acting as microphones for vibrations and changing their output mode pattern with small temperature fluctuations. We find in our laboratory that even with maximal precautions, the output of such a single-mode fiber still has beam-pointing fluctuations on the order of 10 μrad. Depending on the specific experimental situation, this can be unacceptable. Fibers are also costly, produce coupling losses, and need careful alignment. Therefore, fibers do not always solve the problem. Active feedback-controlled beam-pointing stabilization is possible and may well be the best way to increase resolution for slow processes (Grafstrom *et al.,* 1988; Siders *et al.,* 1994).

Care also must be taken to prevent additional beam-pointing noise by beam-steering devices that are used to move the laser trap into the field of view of the microscope. Galvanometer mirrors, for example, exhibit thermal jitter in the range of 10 to 100 μrad. If acousto-optic modulators (AOM) are used to control beam pointing, frequency stability of the controller is crucial. Piezoelectrically actuated mirrors show creep and hysteresis effects that need to be controlled by feedback circuitry.

Finally there are always vibrations and drifts in the microscope, in the laser, and in the detection setup. Building vibrations can be cut off by using an optical bench on vibration-isolated supports. These devices eliminate vibrations faster than a few Hz, but still let slow vibrations pass. Acoustic vibrations are also coupled through the air, making it necessary to eliminate strong noise sources.

VI. Conclusions

We have provided a basic tutorial on noise issues in micromechanical experiments that should be helpful for the nonspecialist in designing experiments. Single-molecule experiments are difficult, and it may save a lot of time to be aware of fundamental facts as well as tricks of the trade that can often be unexpected and counterintuitive. Power spectral analysis is a powerful method much used in physics, but often not appreciated in biological applications. There are some universal recipes to reduce noise in micromechanical experiments, such as low-pass filtering and reducing viscous drag on the probe, but probe stiffness does not play a direct role. For fast motions, viscous drag forces on the probe need to be taken into account, with the consequence that a true constant force experiment is not possible in principle. Finally, we have presented a selection of instrumental design criteria that should be of particular relevance to quantitative optical trapping experiments.

Acknowledgements

We acknowledge detailed discussions with Winfried Denk, who first pointed out that force resolution is independent of probe stiffness in micromechanical measurements, as well as with Karel Svoboda and Winfield Hill. We thank Winfried Denk, Winfield Hill, Karel Svoboda, and Manfred Radmacher for their comments on the manuscript. We acknowledge support from the National Science Foundation (grant #BIR-9512699), the Whitaker Foundation, and the donors of the Petroleum Research Foundation, administered by the American Chemical Society.

References

Ashkin, A. (1992). Forces of a single-beam gradient laser trap on a dielectric sphere in the ray optics regime. *Biophys. J.* **61,** 569–582.

Ashkin, A., and Gordon, J. P. (1983). Stability of radiation-pressure particle traps: An optical Earnshaw theorem. *Optics Lett.* **8,** 511–513.

Ashkin, A., Dziedzik, J. M., Bjorkholm, J. E., and Chu, S. (1986). Observation of a single-beam gradient force optical trap for dielectric particles. *Optics Lett.* **11,** 288–290.

Betzig, E., and Chichester, R. J. (1993). Single molecules observed by near-field scanning optical microscopy. *Science* **262,** 1422–1428.

Bobroff, N. (1986). Position measurement with a resolution and noise-limited instrument. *Rev. Sci. Instrum.* **57,** 1152–1157.

Coppin, C. M., Finer, J. T., Spudich, J. A., and Vale, R. D. (1996). Detection of sub-8-nm movements of kinesin by high-resolution optical-trap microscopy. *Proc. Natl. Acad. Sci. USA* **93,** 1913–1917.

Denk, W., and Webb, W. W. (1990). Optical measurement of picometer displacements of transparent microscopic objects. *Appl. Optics* **29,** 2382–2391.

Funatsu, T., Harada, T., Tokunaga, M., Saito, K., and Yanagida, T. (1995). Imaging of single fluorescent molecules and individual ATP turnovers by single myosin molecules in aqueous solution. *Nature* **374,** 555–559.

Gelles, J., Schnapp, B. J., and Sheetz, M. P. (1988). Tracking kinesin-driven movements with nanometre-scale precision. *Nature* **331,** 450–453.

Gittes, F., Mickey, B., Nettleton, J., and Howard, J. (1993). Flexural rigidity of microtubules and actin filaments measured from thermal fluctuations in shape. *J. Cell Biol.* **120,** 923–934.

Grafstrom, S., Harbarth, U., Kowalski, J., Neumann, R., and Noehte, S. (1988). Fast laser beam position control with submicroradian precision. *Optics Comm.* **65,** 121–126.

Happel, J., and Brenner, H. (1983). "Low Reynolds Number Hydrodynamics: With Special Applications to Particulate Media." 1st ed. (M. Nijhoff, ed.) The Hague, Boston, Hingham, Massachusetts: Kluwer.

Horowitz, P., and Hill, W. (1989). "The Art of Electronics." 2nd ed. Cambridge (England), New York: Cambridge University Press.

Inoue, S. (1986). "Video Microscopy." New York: Plenum Press.

Kuo, S. C., and Sheetz, M. P. (1993). Force of single kinesin molecules measured with optical tweezers. *Science* **260,** 232–234.

Kuo, S. C., Ramanathan, K., and Sorg, B. (1995). Single kinesin molecules stressed with optical tweezers. *Biophys. J.* **68,** 74S.

Landau, L. D., Lifshits, E. M., and Pitaevskii, L. P. (1980). "Statistical Physics." Oxford, New York: Pergamon Press.

Meyhofer, E., and Howard, J. (1995). The force generated by a single kinesin molecule against an elastic load. *Proc. Natl. Acad. Sci. USA,* **92,** 574–578.

Molloy, J. E., Burns, J. E., Sparrow, J. C., Tregear, R. T., Kendrick-Jones, J., and White, D. C. (1995). Single-molecule mechanics of heavy meromyosin and S1 interacting with rabbit or Drosophila actins using optical tweezers. *Biophys. J.* **68,** 298S–303S.

Press, W. H. (1992). "Numerical Recipes in C: The Art of Scientific Computing." 2nd ed. Cambridge, New York: Cambridge University Press.

Radmacher, M., Fritz, M., and Hansma, P. K. (1995). Imaging soft samples with the atomic force microscope: Gelatin in water and propanol. *Biophys. J.* **69,** 264–270.

Reif, E. (1965). "Fundamentals of Statistical and Thermal Physics." New York: McGraw-Hill.

Rugar, D., and Hansma, P. (1990). Atomic force microscopy. *Physics Today* **43,** 23–30.

Sase, I., Miyata, H., Corrie, J. E., Craik, J. S., and Kinosita, K., Jr. (1995). Real-time imaging of single fluorophores on moving actin with an epifluorescence microscope. *Biophys. J.* **69,** 323–328.

Schnapp, B. J., Gelles, J., and Sheetz, M. P. (1988). Nanometer-scale measurements using video light microscopy. *Cell Motil. Cytoskeleton* **10,** 47–53.

Siders, C. W., Gaul, E. W., Downer, M. C., Babine, A., and Stepanov, A. (1994). Self-starting femtosecond pulse generation from a Ti:sapphire laser synchronously pumped by a pointing-stabilized mode-locked Nd:YAG laser. *Rev. Sci. Instrum.* **65,** 3140–3144.

Sigworth, F. J. (1995). Electronic design of the patch clamp. *In* "Single-Channel Recording" (B. Sakman and E. Neher, eds.), 2nd ed., pp. 95–127. New York: Plenum Press.

Simmons, R. M., Finer, J. T., Chu, S., and Spudich, J. A. (1996). Quantitative measurements of force and displacement using an optical trap. *Biophys. J.* **70,** 1813–1822.

Smith, S. B., Cui, Y. J., and Bustamante, C. (1996). Overstretching B-DNA—the elastic response of individual double-stranded and single-stranded DNA molecules. *Science* **271,** 795–799.

Svoboda, K., and Block, S. M. (1994a). Biological applications of optical forces. *Annu. Rev. Biophys. Biomol. Struct.* **23,** 247–285.

Svoboda, K., and Block, S. M. (1994b). Force and velocity measured for single kinesin molecules. *Cell* **77,** 773–784.

Svoboda, K., Schmidt, C. F., Schnapp, B. J., and Block, S. M. (1993). Direct observation of kinesin stepping by optical trapping interferometry. *Nature* **365,** 721–727.

Thomson, N. H., Fritz, M., Radmacher, M., Cleveland, J. P., Schmidt, C. F., and Hansma, P. K. (1996). Protein tracking and detection of protein motion using atomic force microscopy. *Biophys. J.* **70,** 2421–2431.

Wax, N. (1954). "Selected Papers on Noise and Stochastic Processes." New York: Dover.

Yin, H., Wang, M. D., Svoboda, K., Landick, R., Block, S. M., and Gelles, J. (1995). Transcription against an applied force. *Science* **270,** 1653–1657.

CHAPTER 9

Cell Membrane Mechanics

Jianwu Dai and Michael P. Sheetz
Department of Cell Biology
Duke University Medical Center
Durham, North Carolina 27710

I. Perspectives and Overview

Biological membranes are critical to the life of a cell because the hydrophobic interior of the phospholipid bilayer forms a barrier to the transport of solutes and macromolecules between the cell interior and its environment. Furthermore, we are now starting to appreciate the fact that the mechanical properties of these membranes can be utilized by the cell to modify a number of important cell functions. For example, transport of ions and small molecules can occur via specialized proteins in the cell membrane, but a number of transport processes (endocytosis, exocytosis, and other membrane fusion events) involve large, local deformations of the bilayer itself. Indeed, many cell phenomena are accompanied by morphological changes in the cells and so can be affected by the intrinsic

Copyright © 1998 by Academic Press. All rights of reproduction in any form reserved.
0091-679X/98 $25.00

deformability of the membrane and tension within it. Thus, the mechanical properties of cell membrane are involved in a variety of cellular processes at a fundamental level and place physical constraints on such functions. Laser tweezers are ideally suited for analyzing these mechanical properties of plasma membranes in intact cells and of internal membranes *in vitro*.

In many respects biological membranes can be treated as two-dimensional materials. Strictly speaking, they are continuous only in the two dimensions of the cell surface, with a molecular character in the direction normal to the surface. The important mechanical properties of the membrane include its elastic modulus, shear modulus, bending stiffness, and viscosity. From the view of mechanics, membranes have a remarkably low shear modulus (a result of the fluid nature of the lipids), a high elastic modulus (indicative of the lack of stretch in bilayers), and a reasonable bending stiffness influenced strongly by the membrane proteins, including cytoskeletal elements. Another mechanical feature of bilayers is that the two surfaces can be independently modulated, which will induce curvature through the bilayer couple (Evans, 1974; Evans *et al.,* 1976; Hochmuth and Mohandas, 1972). Although we treat the membrane as a homogeneous fluid structure, there are definite domain differences that are being probed with laser tweezers (see Kusumi, this volume). Consequently, membranes are appropriately represented as two-dimensional continua, with possibly an isotropy in the surface plane.

Experiments to determine mechanical properties of biological membranes were begun in the 1930s using sea urchin eggs and, subsequently, nucleated red blood cells (Cole, 1932; Norris, 1939). The early experimentalists concluded that the typical cell membrane is a composite material made up of two molecular layers of lipids plus additional materials presumed to be proteins (Norris, 1939). Experimentally, the determination of material properties of biological membranes (e.g., elastic moduli and viscosity coefficients) involves applying prescribed mechanical forces and observing the resulting change in the shape of the membrane and the time rate of change of membrane conformation. Several other experimental techniques have been used to investigate the mechanical character of cell membranes (e.g., the micropipet aspiration technique and the compression of the cell with these two plates to deform the cells) (Cole, 1932; Mitchison and Swann, 1954).

Figure 1 illustrates these two often-used techniques to deform cell membranes and to study their material properties. Using these methods, the membrane mechanical properties of lipid vesicles, urchin eggs, and red blood cells have been studied extensively (Bo and Waugh, 1989; Cole, 1932; Evans, 1980; Evans and Yeung, 1994; Hochmuth *et al.,* 1973; Norris, 1939). Unfortunately, these techniques are applicable mainly to suspension cells with simple morphology and are inapplicable to cells with a complex structure such as neuronal cells.

The interpretation of the membrane contribution in many such measurements is complicated by the fact that the cytoskeleton is also deformed in a major way. To circumvent the direct cytoskeletal contribution, highly curved membrane

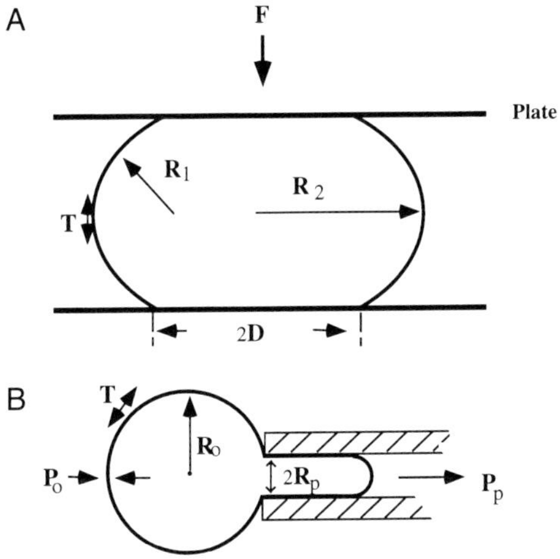

Fig. 1 Diagrams of mechanical deformation of the plasma membrane of a spherical living cell or a lipid vesicle. (A) The cell or vesicle was compressed with known force F between two parallel plates. R_1 and R_2 are the radii of the principal curvatures of the surface. The applied force F divided by the contacted area ($A = \pi D^2$) between the plate and the cell or vesicle (F/A) is the internal pressure P. The pressure is in equilibrium with surface tension T. The surface tension T can be calculated by this equation: $T = P/(1/R_1 + 1/R_2) = F/A(1/R_1 + 1/R_2)$. (B) The diagram of the membrane of a spherical cell or vesicle deformed by a micropipet. P_p is the pressure in the pipet and P_o is the pressure in the reservoir. The isotropic stress resultant in the membrane is the surface tension T, and it is determined by the following equation: $T = (P_o - P_p)/2(1/R_p - 1/R_o)$. R_o and R_p are the radii of the cell or vesicle and the pipet.

cylinders, called tethers, which lack a continuous cytoskeleton, have been studied in red blood cells and pure lipid bilayers with micropipet techniques (Fig. 2). From tether experiments, the static and dynamic components of the membrane

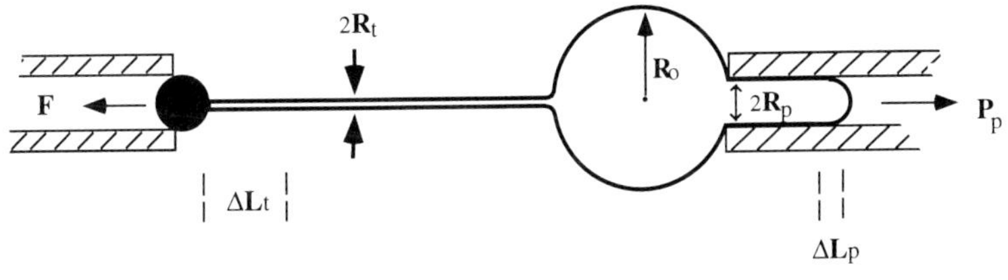

Fig. 2 Two micropipets were used for extraction of a membrane tether from a cell or vesicle. A larger pipet (~10-μm diameter) was used to hold the cell or vesicle with a suction pressure, and a smaller pipet (~4–5-μm diameter) was used to form the tether with a bead attached to the membrane surface with a suction force F. F is the force to form a tether. The tether radius R_t can be calculated from the change in the length of membrane projection in the pipet (ΔL_p) caused by the tether length change (ΔL_t) with the following equation: $R_t = R_p(1 - R_p/R_o)(\Delta L_p/\Delta L_t)$. R_o and R_p are the radii of the cell or vesicle and the pipet.

mechanical properties have been determined. The static tension on tethers contains contributions from the in-plane tension; from the membrane-bending stiffness, which is highly curved in the tethers; and from the membrane–cytoskeleton interaction (Hochmuth *et al.,* 1996; Sheetz and Dai, 1996). When the tether is elongated, a viscous force is introduced that contains contributions from the membrane viscosity, membrane-cytoskeleton interactions, and the interbilayer shear as the lipids flow onto the tethers. The fluid nature of such tethers indicates that they are largely membranous, and the absence of spectrin or actin in erythrocyte tethers has shown that even the membrane cytoskeleton is depleted (Berk and Hochmuth, 1992).

Through tether formation with the micropipet, many mechanical properties (such as the membrane tension, shear rigidity, membrane viscosity, and even membrane thickness) of red blood cell membrane have been determined. Thus, membrane tether formation is a very valuable way to study membrane mechanical properties.

II. Laser Optical Tweezers

Because the existing methods of analysis are not applicable to cells with complex morphology, the studies of membrane mechanical properties generally have been limited to a few kinds of cells with simple morphology, such as red blood cells and neutrophils. However, most cells do not grow in suspension, and most cells have a complex morphology. The advent of laser optical tweezers provides a very flexible method for measuring the cell membrane mechanical properties by tether formation (Dai and Sheetz, 1995a, 1995b, 1995c; Sheetz and Dai, 1996; Hochmuth *et al.,* 1996). Recently, we have extended these studies to determine the membrane mechanical properties by applying forces on beads that bind to membrane surface and by forming membrane tethers with laser tweezers (Dai and Sheetz, 1995a). We now discuss this technique in detail.

III. Bead Coating

To form a membrane tether, a "handle" is required for the laser tweezers to grab a part of the membrane. The handle is typically a latex bead in a range of sizes (0.2 to 2 μm in diameter) that tightly binds to the cell's surface. The bead can be coated with different proteins such as antibodies [(i.e., immunoglobulin G (IgG), lectins (i.e., concanavalin A), extracellular matrix (fibronectin, laminin)], and so forth. The bead can be coated with the relevant molecules in many ways including noncovalent adsorption, direct covalent linkage, or indirect linkage through protein A or an antibody. Following is one of the methods often used to coat beads.

A. Bead Coating Protocol

This is the protocol that we have used to coat 0.2-, 0.5-, or 1-μm beads with rat IgG. It should be noted that salt and protein concentrations often are modified for different bead sizes and different proteins.

Materials: carboxylate spheres (Polysciences, Warrington, PA) or latex beads (Duke Scientific, Palo Alto, CA) and rat IgG (Sigma, St. Louis, MO).

1. Take 100 μl beads from the stock solution and sonicate them (probe at 10 W for 10–20 sec).
2. Take 100 μl coating protein solution (~0.1 mg/ml in PBS) and mix with the beads.
3. Incubate for ~1.5 hr at room temperature or overnight at 4°C.
4. Centrifuge at 10,000 g for 10 min and remove supernatant.
5. Resuspend the beads in 1.5-ml PBS containing 2% BSA. Sonicate for ~30 sec to resuspend the beads completely.
6. Centrifuge at 10,000 g for 10 min and wash the beads three times.
7. Resuspend in 100-μl PBS and store at 4°C for use.

The coated beads can be kept for use up to 1 week.

B. Potential Problems

1. *Bead aggregation.* Often smaller (<0.2 μm) beads will aggregate when salt is added. If the protein can be added at a low-salt concentration, it will coat the beads and prevent salt-induced aggregation.
2. *Weak binding to cells.* Different cell types have different binding affinities for the same coated beads. A balance between binding and activation is needed. With an affinity that is too low, binding will not occur. With an affinity that is too high or with specific receptor binding, activation of cell motility or cytoskeleton binding can occur.

IV. Calibration of the Laser Tweezers

Presently there is no theory that can be used to directly calculate the trapping force for beads. All the forces must be determined empirically, and the forces are commonly calibrated against viscous drag exerted by fluid flow. For a bead of radius r, the drag force can be obtained from Stokes' Law: $F = 6\pi\,\eta r v$. Here η is the fluid viscosity and v is the flow rate. Following are the two most commonly used ways to calibrate a laser trap.

A. Maximal Force Calibration

The preceding calibration enables knowing what is the maximal force that the trap produces on a certain bead at a certain power output. This calibration is the simplest and can be made very quickly. The viscous force can be produced either through a flow chamber through which fluid is pumped past the stationary, trapped bead at an adjustable velocity or by moving the trapped bead in the stationary fluid with a motorized or piezo-driven stage. The local fluid velocity can be measured by tracking other beads in the flow field in the same focal plane; velocity produced by the piezo-electronic driven stage can be measured by tracking other beads that have settled down on the glass surface. At a certain laser power output, we can increase the viscous force on the bead until the bead escapes the trap and thus can obtain the maximal force the laser trap can produce at this certain power output. The linear plot of the power output via escaping force gives the calibration of the maximal force that can be used to compute the maximal force on a bead of the same size at any power level (Fig. 3).

B. Calibration of Trap by Measuring Trap Stiffness

The force on the bead also can be obtained if we know the laser trap stiffness, because the force can be determined as a function of a displacement from the trap center. To measure the trap stiffness, a viscous force is generated by oscillatory motion of the specimen by a piezoceramic-driven stage at a constant velocity. The position of the bead in the trap is tracked using the nanometer-level tracking program to analyze video records of the experiments. The viscous force on the bead can be calculated through Stokes' Law. The calibration shows a very linear force-displacement relationship for the optical tweezers (Fig. 4); the slope of the linear fit gives the trap stiffness. If the trap stiffness and the displacement of the particle position from the trap center are known, the force on the trap can easily be calculated.

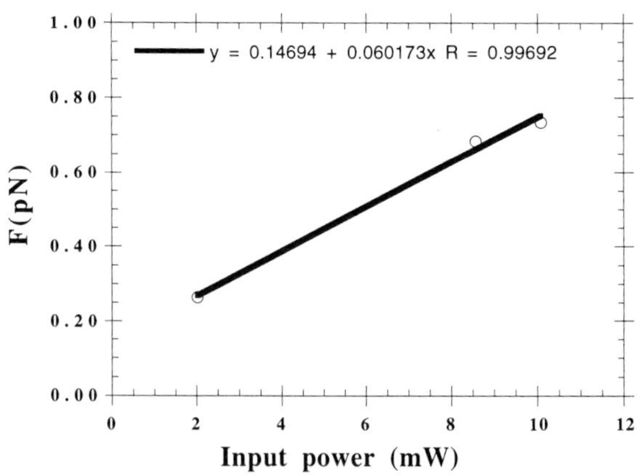

Fig. 3 The calibration of a laser trap by measuring the bead-escaping force from the trap. This calibration can give the maximal force of the trap at a certain power.

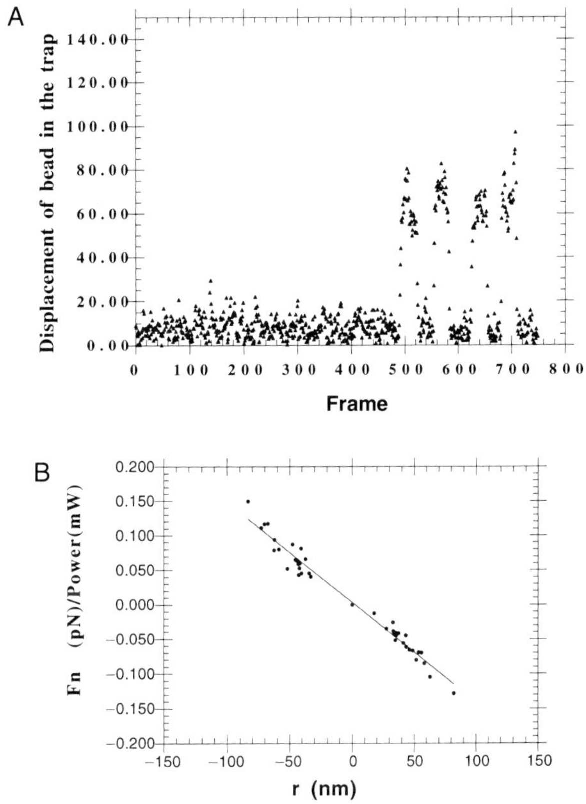

Fig. 4 Calibration of a laser trap by measuring the trap stiffness. (A) An example of the radial position of the bead r in the trap during a series of calibration displacements. The bead was held with the laser trap ~4 μm above the coverslip surface, and the piezoceramic-driven stage was moved repeatedly at a constant rate. (B) Calibration of the laser trap for particles held 3 to ~4 μm above the coverslip surface by varying the laser power for a constant velocity of movement. The result is shown as the relationship between the position of the bead in the trap r and the normalized light force F_N in pN · mW^{-1} as calculated from Stokes' Law and the power of the trap, which was monitored simultaneously. The slope of the linear fit line is the trap stiffness, which is 1.443 (pN · mW^{-1} · μm^{-1}). Reproduced from Fig. 1 from Dai and Sheetz, 1995a, with permission from the Biophysical Society.

We now know how to calibrate the laser trap and accomplish the force measurement, but we should keep in mind that the results of calibration and the force measurement sometimes can be greatly affected by the different perpendicular positions. To characterize the variation in trap calibration with height above the coverslip surface, a bead is trapped with the same laser power at different perpendicular positions, and a certain viscous force is introduced to the bead.

The increase in particle displacement at 2 μm or less from the glass surface (Fig. 5) implies a viscous coupling to the coverslip surface. From 2–5 μm above the surface, the force on the beads in the trap was constant. This indicates that if

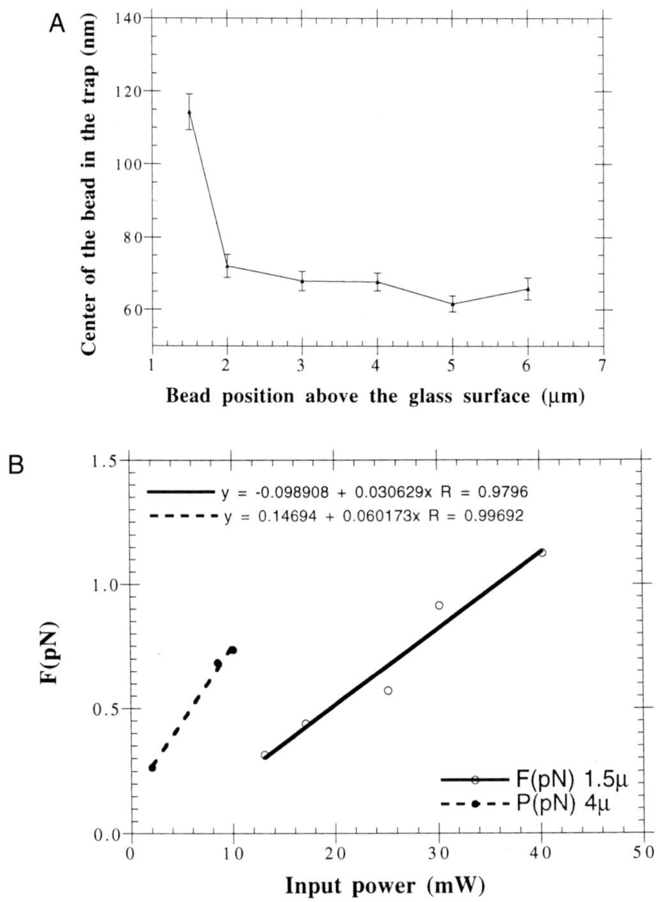

Fig. 5 The effect of distance from the coverslip on the calibrations. The results show that experiments need to be done at least 2 μm above the coverslip surface to minimize viscous coupling to the glass surface. (A) Calibration of the laser trap at different heights above the glass surface. The displacement procedure in Fig. 4A was repeated for latex beads (0.5-μm diameter) held at different heights above the coverslip at constant power. Six beads were tested at each height. (B) Calibrations of a trap by measuring the escaping forces. There was dramatic difference of the calibrations when the calibrations were performed at 1.5 μm or 4 μm above the coverslip surface. Fig. 5A is reproduced from Fig. 1 from Dai and Sheetz, 1995a, with permission from the Biophysical Society.

we are trying to do high-speed stretch, we need to perform the experiment 3–4 μm above the coverslip surface to minimize viscous coupling to the glass surface.

V. Tracking the Bead Position

Tracking the bead position with nanometer-level resolution is important for both the calibration of trap stiffness and the tether force measurement. In our laboratory, in addition to using the Macintosh-LVR500-LG3 tracking station (analysis software is NIH Image V1.59) and the Image-1/MetaMorph Imaging System (Universal Imaging Co., West Chester, PA), we mainly use the IBM-Driven Analysis Station. Two programs (FLOX and FLOC) are used for tracking the bead. The difference between the two is in their method of tracking: The fast locate correlation (FLOC) program simply locates a bead position by finding an intensity peak in a defined area, a region of interest (ROI). Then using a standard weighted centroid calculation for the pixels higher than a defined threshold, the program geometrically determines a centroid of the object (bead or any other particles) and records this as the bead position. The fast locate xross correlation (FLOX) is a more accurate method of tracking. A kernel image of the bead is saved to a disk from an RTD Box program and is used as a template for the program to locate this same image in the subsequent frames within the ROI. As the position of the bead moves, the FLOX program automatically moves the ROI to follow the target. The procedures for using this system for tracking bead position are as follows:

1. Digitize and record the interesting sequence on the Real Time Disk using the RTD Box program.
2. Specify ROI by drawing a box around the target bead using the RTD Box.
3. Save parameters about the video sequence into a text file, which can be read by the tracking programs.
4. Save a kernel image of the bead to the hard disk.
5. Run the tracking program FLOX or FLOC, which actually finds the bead position and records it to an ASCII text file that contains position information about the bead or particle tracked.

A. Potential Problems and Solution

Our current tracking systems have significant noise sources including thermal noise, shot noise, and electronic noise. The noise level from the laser and other electronics systems in the room can be as high as 10% of the signal. We are trying to build the quadrant detector system to enable faster and more accurate tracking of moving particles.

VI. Membrane Tether Formation and Tether Force Measurement

To form a membrane tether, the coated beads need to be added to a flow chamber containing the cells or to the medium before the sample is mounted on the stage. Cells must be either grown on the glass surface or attached to the glass surface by some treatments. Then a bead is trapped with the laser tweezers and placed on the cell surface. The bead is held on the surface for a few seconds, then pulled at a constant velocity with a motorized or piezo-driven stage to form a membrane tether (Fig. 6). Different length tethers can be formed (from several to more than 100 μm depending on the cell type). After a tether is formed, it can be kept at constant length for minutes before it breaks or is resorbed. The retraction force of the tether can be obtained by calculating the force on the bead. To analyze the force, the bead position in the trap is tracked off line by a video-based centroid method and, with the laser trap calibration, the force is computed. Alternatively, the bead position in the trap can be measured by imaging the bead onto a split photodiode detector: The difference voltage is proportional to displacement for motions up to the order of the particle radius. The detector needs to be aligned with the laser tweezers. An alternative way is to use an optical trapping interferometer, in which the same laser light serves to produce both the trapping and interferometer functions. Displacement is proportional to the ellipticity developed by polarized light after the interferometer beams are recombined. In this setup, with the trap calibration, the tether force can be measured immediately on line (Svoboda and Block, 1994).

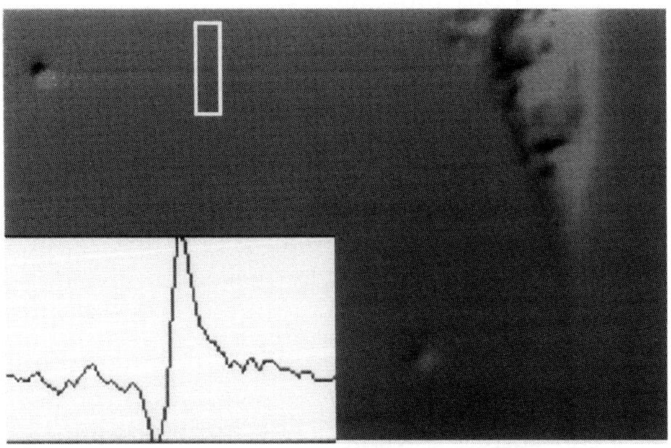

Fig. 6 Photomicrograph of a tether from a living cell. The diagram shows how to determine the relative intensity from the DIC image through orthogonal scans across the DIC image of the tether. The intensity is related to the tether thickness.

VII. Membrane Tether Force and Membrane Mechanical Properties

Laser optical tweezers provide a very flexible way to study membrane mechanics by membrane tether formation. A typical force pattern on the bead during the tether formation includes three different phases (Fig. 7). The first phase gives the force on the bead before the tether starts to elongate. Theoretically, at this phase the tether force on the bead is zero. In fact, this phase gives us the very useful information about system errors, including the trap position stability. If the whole system is stably mounted, this error is about 5% (or <10%) of the tether force measured.

The second phase gives the force during tether elongation. This tether force F contains both the static and dynamic membrane mechanical components. Plots of F as a function of tether elongation rate (V) are linear with a slope proportional to an apparent viscosity and an intercept at zero velocity that is the static tether force (F_0). F_0 can also be obtained from the static phase of the measurement. F_0 contains only the static membrane mechanical components (i.e., membrane in-plane tension, bending stiffness of the bilayer, and membrane–cytoskeleton interaction). In all cell systems studied so far, we have not found a significant change in F_0 with tether length more than what we refer to as a plateau phase. When tethers are further elongated, the tether force rises exponentially with length (Raucher and Sheetz, unpublished results).

With F and F_0 measured directly by laser optical tweezers, many other mechanical parameters can be calculated. We give some examples here.

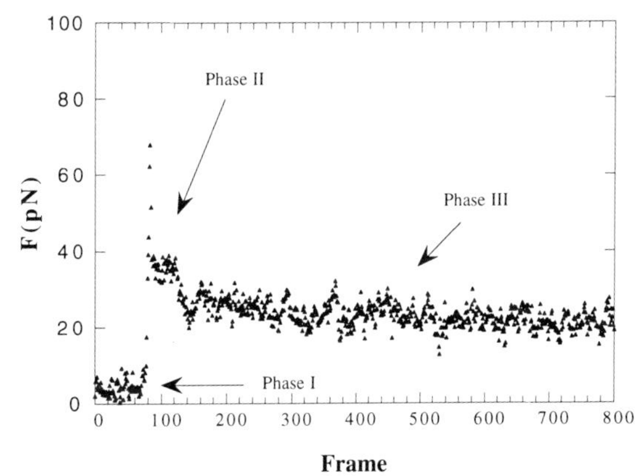

Fig. 7 The tether force pattern when a tether was formed from a rat basophilic leukemia (RBL) cell. It can be defined into three phases: phase I (before starting to pull the bead), phase II (during the pulling or the tether elongation), and phase III (when the tether is kept at a constant length).

The first parameter that can be instantly calculated is the apparent membrane viscosity. The viscosity of a biological membrane is an important factor in determining the rate at which the membrane can undergo deformation on a microscopic scale, and it also influences the rate at which membrane components such as integrin molecules and receptor molecules can diffuse in the plane of the surface. These processes play important roles in a wide variety of biological phenomena. If we have the experimentally measurable parameters, F, F_0 and V, a theory developed by Waugh (Waugh, 1982) can be applied to compute membrane surface viscosity. Using this method, we obtained the viscosity for the chick dorsal root ganglion (DRG) neuronal growth cone membrane. This viscosity is in the range of 0.56×10^{-4} to 4.2×10^{-4} dyne · sec/cm (Dai and Sheetz, 1995a). From the studies of pure lipid membranes and the effects of cytoskeletal perturbing drugs, it is evident that the major viscous component is not the lipid viscosity but rather the drag of the lipid across the membrane skeleton.

Another mechanical parameter readily calculated is the cell membrane bending stiffness. This parameter characterizes the properties of the membrane bilayer itself; its large resistance to area changes results in a small but finite resistance to changes in curvature. The bending stiffness B can be calculated from measurements of tether radius R_t as a function of static force F_0 on the tether (Bo and Waugh, 1989; Hochmuth *et al.*, 1996): $B = F_0 \cdot R_t/2\pi$. Currently there are two ways to measure the tether radius. A known-size particle is used as a standard. Using certain computer programs, such as Ruler in our laboratory, from the videotape we can measure the relative diameter of the tether and then compare it with the bead. The other way is to determine the relative intensity through orthogonal scans across the differential interference contrast (DIC) image of both the tether and some cell structures, such as an axon (Fig. 6). After measuring the diameter of the structure (e.g., the diameter of the axon) with the Ruler program, we can estimate the radius of the tether: The intensity should be related to the radius (Hochmuth *et al.*, 1996; Schnapp, *et al.*, 1988). In the future, the scanning electronic microscope (SEM) will be used for accurate measurement of tether radius. The radius of the tether from the chick DRG growth cone has been calculated to be approximately 0.2 μm in both cases. From the preceding equation, the bending stiffness we obtained for the growth cone membrane is 2.7×10^{-19} N · m (Hochmuth *et al.*, 1996).

The third example of membrane mechanical parameters that can be calculated from experimentally measurable parameters in the tether formation experiments with laser optical tweezers is the membrane tension. Because we feel that this parameter is more complicated and of special importance to many cell functions, we discuss it in a separate section.

VIII. Membrane Tension and Its Significance

Tension within plasma membrane could regulate many important cell activities such as membrane trafficking (i.e., exo- and endocytosis), cell division, and cell motility because all these cell phenomenon may involve membrane tension changes. The technique of tether formation with laser optical tweezers provides a unique way to measure membrane tension and to enable us to better understand the mechanisms of these fundamental cell processes. The membrane tension term of the static tether force F_0 is complicated for cultured cells because it contains the contributions from in-plane membrane tension, membrane–cytoskeleton adhesion and membrane-bending stiffness B. In terms of membrane tension, the human erythrocyte and lipid vesicle are two extremes. In the human erythrocyte the in-plane tension is very small; in the lipid vesicle the membrane–cytoskeleton is missing. In lipid vesicles, the in-plane tension can easily be adjusted by applying a known suction pressure to the vesicle with a micropipet. In contrast, for most eukaryotic cells, it is difficult to impose a known in-plane tension because the stress applied by the micropipet is primarily absorbed by the cytoskeleton. The tension contributed by the membrane–cytoskeleton interaction is more difficult to understand because of its rapid reversibility. At present, in-plane membrane tension cannot generally be separated from the membrane–cytoskeleton interaction contribution to the tether force because most animal cells adopt complex shapes. These cells have a very complicated cytoskeleton system, and the surface contains folds and invaginations. If the in-plane tension exceeds the adhesive force between the membrane and the cytoskeleton, then the membrane will pull away from portions of the cytoskeleton. This may explain the very small change of tether force with tether length that has been observed experimentally (Dai and Sheetz, 1995). However, this indicates that the in-plane membrane tension and the membrane–cytoskeleton interaction are linked. Consequently, we have combined both into a single membrane tension term, T_m. The relationship between the static tether force (F_0), bending stiffness (B), the radius of the tether (R_t) and membrane tension (T_m) is given as follows (Hochmuth *et al.,* 1996; Sheetz and Dai, 1996):

$$F_0 = \pi B/R_t + 2\pi R_t T_m \tag{1}$$

If we combine Eq. (1) with Eq. (2) that we have used to calculate bending stiffness, either the R_t term or the B term can be removed and we obtain Eq. (3) and (4):

$$B = F_0 R_t / 2\pi \tag{2}$$

$$F_0 = 2\pi \sqrt{2BT_m} \tag{3}$$

$$F_0 = 4\pi R_t T_m \tag{4}$$

Using the preceding equations, if we can experimentally measure any two of the four parameters in the equations, then the other two can be calculated. This enables cell membrane tension to be determined under widely different conditions by tether formation with laser optical tweezers.

In interpreting the possible effects of membrane tension on cell functions, we studied the effects of membrane tension on the rat basophilic leukemia (RBL) cell-membrane-trafficking processes, exocytosis and endocytosis. We studied the membrane tension of the cell under different conditions, which included altering tension by osmotic swelling and stimulating cell secretion by cross-linking cell immunoglobulin (IgE) receptors. The observations indicate that there is a strong correlation between changes in membrane tension and the rates of exocytosis and endocytosis (Dai *et al.,* 1997). The observations also support the hypothesis that the membrane tension may play a major role in controlling membrane trafficking, cell shape, and motility (Sheetz and Dai, 1996). In emphasizing the importance of membrane tension, it is reasonable to speculate that membrane tension also may control cell mitosis, cell death, and many other cell functions because cells experience great shape changes in these processes.

The laser tweezers provide us a great opportunity to study cell membrane mechanical properties and to understand better the mechanisms of some fundamental cell functions. This has already led us to a totally new field with a promising future.

References

Berk, D. A., and Hochmuth, R. M. (1992). Lateral mobility of integral proteins in red blood cell tethers. *Biophys. J.* **61,** 9–18.

Bo, L., and Waugh, R. E. (1989). Determination of bilayer membrane bending stiffness by tether formation from giant, thin-walled vesicles. *Biophys. J.* **55,** 509–517.

Cole, K. S. (1932). Surface force of the *Arbacia* egg. *J. Cell. Comp. Physiol.* **1,** 1–9.

Dai, J., Ting-Beall, H. P., and Sheetz, M. P. (1997). The secretion-coupled endocytosis correlates with membrane tension changes in RBL $2H_3$ cells. *J. Gen. Physiol.* **110,** 1–10.

Dai, J., and Sheetz, M. P. (1995a). Mechanical properties of neuronal growth cone membranes studied by tether formation with laser optical tweezers. *Biophys. J.* **68,** 988–996.

Dai, J., and Sheetz, M. P. (1995b). Axon membrane flows from the growth cone to the cell body. *Cell* **83,** 693–701.

Dai, J., and Sheetz, M. P. (1995c). Regulation of endocytosis, exocytosis, and shape by membrane tension. *In* "Protein Kinesis: Dynamics of Protein Trafficking and Stability,"pp. 567–571. Coldspring Harbor, New York: Cold Spring Harbor Laboratory Press.

Evans, E. A. (1974). Bending resistance and chemically induced moments in membrane bilayers. *Biophys. J.* **14,** 923–931.

Evans, E. A. (1980). Minimum energy analysis of membrane deformation applied to pipet aspiration and surface adhesion of red blood cells. *Biophys. J.* **30,** 265–284.

Evans, E. A., and Yeung, A. (1994). Hidden dynamics in rapid changes of bilayer shape. *Chem. Phys. Lipids* **73,** 39–56.

Evans, E. A., Waugh, R., and Melink, L. (1976). Elastic area compressibility modulus of red cell membrane. *Biophys. J.* **16,** 585–595.

Hochmuth, R. M., and Mohandas, N. (1972). Uniaxal loading of the red cell membrane. *J. Biomech.* **5,** 501–509.

Hochmuth, R. M., Mohandas, N., and Blackshear, P. L. (1973). Measurement of the elastic modulus for red cell membrane using a fluid mechanical technique. *Biophys. J.* **13,** 747–762.

Hochmuth, R. M., Shao, J., Dai, J., and Sheetz, M. P. (1996). Deformation and flow of membrane into tethers extracted from neuronal growth cones. *Biophys. J.* **70,** 358–369.

Mitchison, J. M., and Swann, M. M. (1954). The mechanical properties of the cell surface: I. The cell elastimeter. *J. Exp. Biol.* **31,** 443–460.

Norris, C. H. (1939). The tension at the surface, and other physical properties of the nucleated erythrocyte. *J. Cell Comp. Physiol.* **4,** 117–128.

Schnapp, B. J., Gelles, J., and Sheetz, M. P. (1988). Nanometer-scale measurements using video light microscope. *Cell Motil. Cytoskeleton* **10,** 47–53.

Sheetz, M. P., and Dai, J. (1996). Modulation of membrane dynamics and cell motility by membrane tension. *Trends Cell Biol.* **6,** 85–89.

Svoboda, K., and Block, S. M. (1994). Biological applications of optical forces. *Annu. Rev. Biophys. Biomol. Struct.* **23,** 247–285.

Waugh, R. E. (1982). Surface viscosity measurements from large bilayer vesicle tether formation. I. Analysis. *Biophys. J.* **38,** 19–27.

CHAPTER 10

Application of Laser Tweezers to Studies of the Fences and Tethers of the Membrane Skeleton that Regulate the Movements of Plasma Membrane Proteins

Akihiro Kusumi,* Yasushi Sako,* Takahiro Fujiwara,† and Michio Tomishige†

* Department of Biological Science
Graduate School of Science
Nagoya University
Nagoya 464-01, Japan

† Department of Life Sciences
Graduate School of Arts and Sciences
The University of Tokyo
Komaba 3-8-1, Meguro-ku
Toyko 153, Japan

Copyright © 1998 by Academic Press. All rights of reproduction in any form reserved.
0091-679X/98 $25.00

I. Introduction

Our understanding of the cellular mechanisms that control the movement and assembly of membrane proteins in the plasma membrane is currently undergoing rapid evolution. It is becoming clear that movement and assembly are, in part, regulated through the membrane-associated portion of the cytoskeleton (i.e., the membrane skeleton) (Bennett, 1990; Bennett and Gilligan, 1993; Edidin, 1992, 1993; Jacobson *et al.,* 1995; Kusumi and Sako, 1996; Luna and Hitt, 1992; Sheets *et al.,* 1995). Of particular interest is the involvement of membrane–skeletal elements in the assembly of receptors for the formation of specialized membrane domains and self-association of cell surface receptors by mediating or inhibiting the movements of the membrane proteins in the plasma membrane.

For the advancement of such studies, newly developed light microscopic techniques, such as single-particle tracking (SPT) and laser tweezers, are making important contributions. In this chapter, the application of laser tweezers to studies of the mechanisms by which cells regulate the movement and lateral diffusion of membrane proteins is covered. In these experiments, gold or latex particles are attached to membrane proteins, and the particle–membrane protein complex is captured by laser tweezers and moved laterally along the plasma membrane by scanning the trapping laser beam (or by moving the stage, Fig. 1). Recent reviews have covered aspects of cytoskeleton–membrane interactions (Hitt and Luna, 1994; Kusumi and Sako, 1996), cytoskeletal control of the lateral diffusion of membrane proteins (Jacobson *et al.,* 1995; Sheets *et al.,* 1995), and the localization of membrane proteins during polarization development (Nelson, 1992).

Two types of major interactions between the membrane skeleton and membrane proteins have been found: One is the binding of the membrane proteins to the membrane skeleton (Fig. 2A), and the other is the corralling or fence

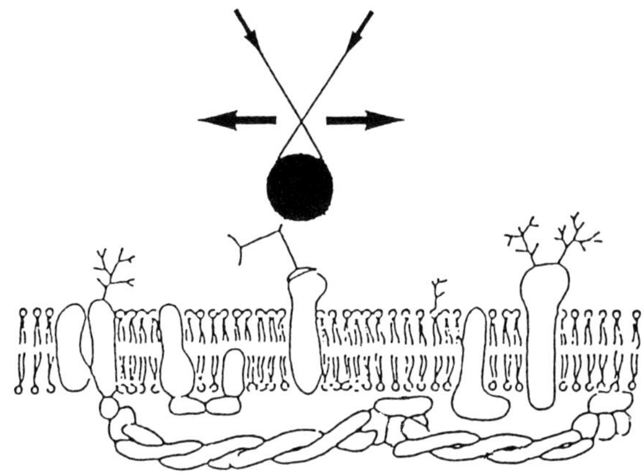

Fig. 1 Conceptual cartoon showing an experiment using laser tweezers to drag a membrane protein molecule along the plasma membrane.

effects of the membrane skeleton on the lateral diffusion of membrane proteins (Fig. 2B). Some membrane proteins (e.g., E-cadherin, which is responsible for calcium-dependent cell-cell adhesion) experience both effects of the membrane skeleton at the same time. By using laser tweezers to move membrane proteins and the membrane skeleton at the level of single molecules, these interactions can be studied. The forces involved in such interactions, at a level of subpico- to pico-newtons (pN), and the elasticity of the membrane skeleton involved in the tethering or corralling of the membrane protein at a level of 1–10 pN/μm can be observed.

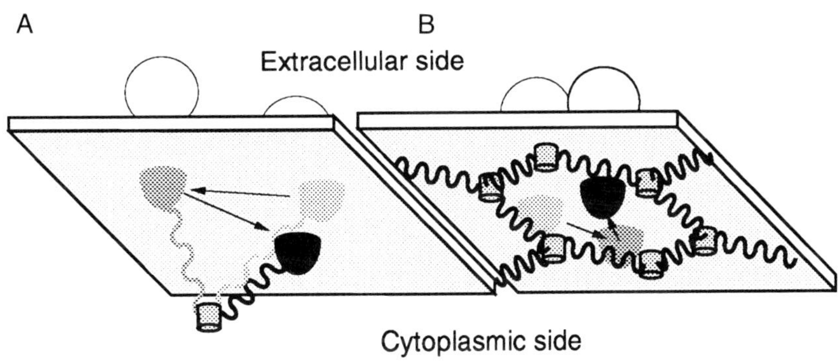

Fig. 2 Schematic models of two types of interactions between membrane proteins and the membrane skeleton. (A) A membrane protein molecule is anchored to the membrane–skeleton/cytoskeleton (tether model). (B) A membrane–protein molecule is confined by the membrane–skeleton network (fence model), in which the movements of the membrane protein are limited by the compartment boundaries consisting of membrane–skeleton network barriers located near the membrane.

Because the application of laser tweezers to studies of barriers for lateral movements of plasma membrane proteins has only recently been attempted, there are few well-established methods. Therefore, we describe several representative experiments that have been carried out recently, and hope that these inspire readers to devise new experiments using laser tweezers to study the mechanics of the molecular interactions and structures in the plasma membrane.

II. Studying Various Modes of Protein Motion in the Plasma Membrane Using Single-Particle Tracking

In single particle tracking (SPT), nanometer-sized colloidal gold (or fluorescent) particles are coated with specific antibodies to membrane proteins, or ligands to receptor molecules, and then are attached to a single (or a small number of) membrane protein molecule(s) via the particle-bound antibodies or the ligands. The methods for preparing these gold particles and binding proteins on the particles have been described (Leunissen and de Mey, 1989; Roth, 1983). The gold-receptor complexes are visualized by contrast-enhanced optical microscopy, and the movements of the complexes can be followed, with the possibility of nanometer-level precision (Anderson *et al.*, 1992; Cherry, 1992; Geerts *et al.*, 1991; Gelles *et al.*, 1988; Ghosh and Webb, 1994; Sheets *et al.*, 1995; Wang *et al.*, 1994).

Kusumi *et al.* (1993), Qian *et al.* (1991), and Saxton (1993, 1994, 1995, 1996) presented the theoretical basis for analyzing the trajectories obtained by SPT. A practical approach for the analysis of single-particle trajectories has been described, enabling the true deviations from Brownian motion to be distinguished from the statistical fluctuations inherent to random walks (Kusumi *et al.*, 1993). Using this approach, Kusumi *et al.* (1993) and Tomishige (1995) found that the modes of motion of many membrane proteins can be classified into the five types of motion believed to occur in the plasma membrane: (a) stationary mode; (b) simple Brownian diffusion mode; (c) directed diffusion mode; (d) confined (or corralled) diffusion mode (in which a particle undergoing free diffusion is confined within a limited area); and (e) diffusion in a potential well, which often can be approximated by a harmonic potential. A single protein molecule can change its motional modes with time (Kusumi *et al.*, 1993; Sheetz *et al.*, 1989; Simson *et al.*, 1995).

III. Technical Details for SPT

SPT is an essential part of laser tweezer experiments with gold or latex probes, because the behaviors of these probes under the trapping forces are observed by SPT.

A. Preparation of Gold Particles for Labeling Membrane Proteins

The standard methods for preparing gold particles and for binding proteins on the practical have been described in detail (Roth, 1983; de Mey, 1983; Leunissen and de Mey, 1989). Therefore, we will simply add several comments that may be practically helpful, based on our past experience (mostly unpublished).

1. In many cases, we observed that we had to add more protein (ligands or antibodies) than the minimal protecting amount (the minimal amount of the proteins required to coat the gold surface to prevent aggregation and sedimentation of the gold particles in salt solutions) to make the coated gold particles specifically attach to the target membrane proteins on the cell surface. This suggests that (a) the proteins tend to denature on the gold surface, and (b) as the concentration of the protein in the conjugation medium increases, the proteins on the gold surface may become more crowded immediately after mixing, and the extra room for the bound protein to be extended on the surface (and denatured) is more limited, which results in functional gold probes.

2. Nevertheless, it is useful to know that a 15 nm-ϕ gold particle can bind approximately 60 protein A or 31 transferrin molecules on its surface (Baudhuin *et al.,* 1989), as a basic figure for thinking about the geometries of gold probes. Table 1.2 in Baudhuin *et al.* (1989) lists the number of molecules/particle and the dissociation constants for various proteins (but not IgG).

3. Only a small portion of the ligand molecules on the gold surface may be functional, because of their orientations with respect to the gold surface.

4. Only small portions of the ligands attached to the gold particle may be simultaneously functional on the cell surface, because only a small part of the gold surface can have access to the cell surface at one time.

5. In many cases, gold particles coated with ligands or IgG′s (or Fab′s) behaved like simple ligands (Mecham *et al.,* 1991; possibly with a bit higher affinity, but not much), that is, coexistence of the ligand proteins effectively competes with their gold conjugates (Kusumi *et al.,* 1993; Sako and Kusumi, 1994). This may be due to the perviously mentioned geometrical reasons of points (3) and (4) above, and also because of the high rate of denaturation of ligands on the gold surface, as described in (1).

6. As proteins that mediate the binding of gold particles to membrane proteins on the cell surface, we prefer using monovalent proteins, that is, monovalent ligands or antibody Fab's rather than IgG's. This may seem odd because there are many bound protein molecules on a gold particle. However, points (1)–(4) suggest that functional ligand molecules tend to be sparse and isolated on the gold surface, and that it makes sense to have these isolated functional ligands monovalent. In some cases, we have seen a great reduction in the mobility when we switched from antibody Fab to IgG (unpublished observations).

7. To make paucivalent gold particles, therefore, it is a good idea to premix specific Fab's with nonspecific Fab's or IgG's to coat the gold surface at various

ratios, and to examine their binding to the cells and their behaviors on the cell surface.

8. The eff rates of Fab or Fab–gold complex tends to fall between 3 and 30 min (unpublished observations).

9. In some cases, we were unable to make functional gold probes simply by adding the ligands to the gold particles, such as in the case of mouse epidermal growth factor (EGF). Functional gold probes for the EGF receptor was made by binding EGF to gold particles via BSA, by using a crosslinker (Kusumi *et al.*, 1993; Yoshitake *et al.*, 1982). First, BSA was attached to the gold particles as described at pH 6.0. The unbound BSA was removed by repeated centrifugation and resuspension of the gold particles. EGF (0.4 mg, HIH, Tokyo), dissolved in 0.3 ml of phosphate buffer (67 mM, pH 7.0), was incubated with 1.6 mg of N-(ε-maleimidecaproyloxy)succinimide (EMCS, Dojin, Kumamoto, Japan) at 30°C for 1 hr. Thc unreacted EMCS was removed by gel filtration on Sephadex G-25 (Pharmacia). The EMCS-bound proteins and the BSA-coated gold particles were mixed and incubated at 4°C for 24 hr.

B. Video Microscopy

The instrumental setup we use for video-enhanced contrast microscopy and image analysis is illustrated in Fig. 3. A Zeiss Axioplan microscope is equipped with a condenser lens (NA = 1.4) and a Plan-Neofluar objective (100×, NA = 1.3; or 63×, NA = 1.). Cells are observed by video-enhanced differential interference contrast microscopy, with illumination through an optical fiber (Technical Video, Woods Hole, MA) by using the green line (wavelength = 546 nm) of a 100-W mercury arc lamp (HBO 100). The image is projected on a Hamamatsu CCD camera (C2400-77, Hamamatsu, Japan). Real-time mottle subtraction and contrast enhancement are achieved by a Hamamatsu DVS-3000 image processor, and the processed image is recorded on a Panasonic TQ-3100F laser disk recorder.

The video images are digitized with the DVS-3000 image processor, and selected areas of the image are sent to a computer (Intel Pentium machine). The positions (x and y coordinates) of selected gold particles are determined automatically with the computer, by using a method developed by Gelles *et al.* (1988). The accuracy of the position measurement was estimated by recording a sequence of 150 video frames of the images of a 40-nm gold particle fixed on a polylysine-coated coverslip and impregnated in a 10% polyacrylamide gel. The standard deviations of the measured coordinates of the fixed particles were 1.8 nm horizontally and 1.4 nm vertically, which were comparable to other published values (0.5–1 nm for 190-nm latex particles, Gelles *et al.*, 1988; Schnapp *et al.*, 1988). The nominal diffusion coefficient of the fixed particle was 3.2×10^{-13} cm^2/sec, which, therefore, is the lower limit for determining the diffusion coefficient with the present settings.

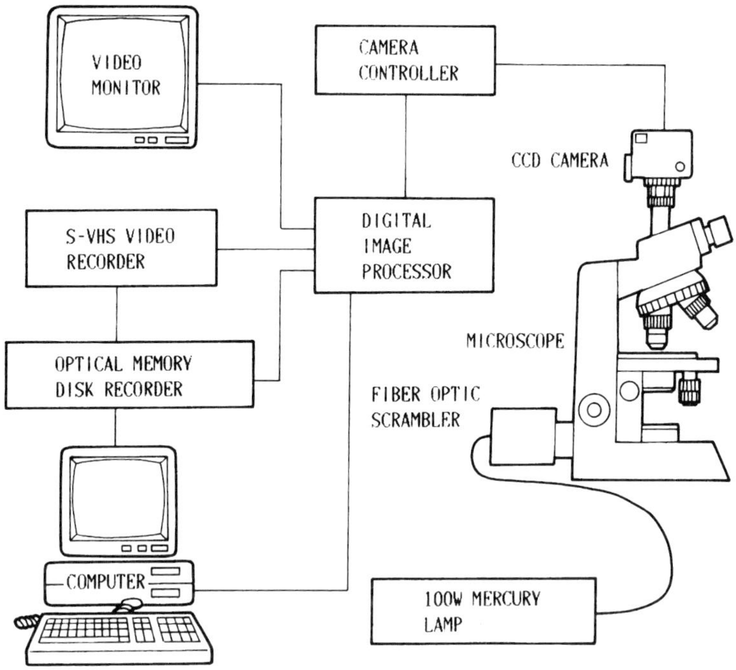

Fig. 3 Block diagram of the instrument used for single particle tracking (nanovid microscopy). See text for details.

IV. Instrumental Setup and Calibration of Laser Tweezers

A. Instrumental Setup of Laser Tweezers

The schematic design of our laser tweezers is shown in Fig. 4. Briefly, a 1063-nm laser beam from a Nd/YAG laser (Amoco, model ALC D500, Naperville, IL, 350 mW output power (380 mW actual power)) is introduced onto a microscope (Zeiss, Axioplan) through an epi-illumination light path. The laser beam is expanded to fill the aperture at the back focal plane of the objective, and is then focused to form a 1-μm spot on the specimen. The maximum incident laser power just before the objective is 310 mW, and that after the objective (100×, Plan Neofluar) is 100 mW. To scan the laser beam in the x and y directions, two mirrors on scanners controlled by a microcomputer are set just before the entrance of the microscope (General Scanning, model 6000 and MmPIC-20A, Watertown, MA). Another design would be to use piezo devices that scan the stage and move the sample to place the optical trap at the desired position. We initially selected the present design because we intended to have two traps under the same microscope to pull the membrane or the membrane skeleton with "two

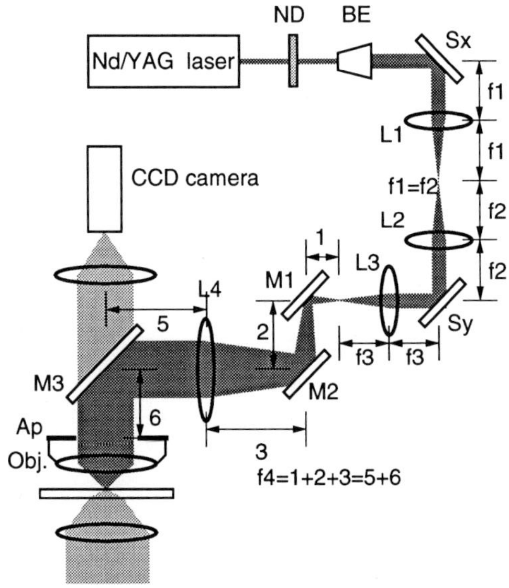

Fig. 4 Block diagram of our laser tweezers system. ND, neutral density filters; BE, beam expander; L, lens; Sx and Sy, scanning mirrors for x and y directions; fn, focal length of the lens Ln (n, integer); M, mirror; 1–6, various distances; Obj, objective lens; Ap, the aperture of the objective lens. f1 = f2. f4 = 1 + 2 + 3 = 5 + 6. The last condition is to scan the beam with the aperture of the objective lens set at the fulcrum of the beam movements.

hands,'' as a future expansion of the instrument, which is now in progress in our laboratory.

B. Simple Calibration of the Maximal Trapping Force of the Laser Tweezers

Latex and gold particles that are conjugated with ligands or antibodies are suspended in sucrose solutions of various concentrations and are dragged by the optical tweezers at laser powers of 100, 50, and 25 mW (measured just after the objective lens). By varying the laser scan speed (v) or the solution viscosity (η) independently, conditions under which half of the particles are dragged for more than 20 μm can be found, and the trapping force (F) under these conditions is determined using Stokes' equation, $F = 6\pi a\eta v$ (a, radius of the particle). In the present setting of the instrument, the scan speed can be varied from 0.6 to 18 μm/sec under a 100× objective. The viscosity of the solution is varied from 0.89 to 43.4×10^{-3} nsec/m^2 (0.89–43.4 cP) by varying the sucrose concentration (the temperature was fixed at 25°C). The forces obtained by varying the scan speed and the viscosity agreed within ± 0.03 pN in all ranges of the trapping force used in this study. However, in general, caution should be exerted for changes of the medium, because they can also change the trapping force for the particles, later particles in particular.

C. Full Calibration of the Force Potential of the Laser Tweezers

Several methods for the calibration of the force potential of the optical trap have been described elsewhere in this book. Our method is based on the measurement of the equilibrium distribution of particles in the optical trap. By measuring the positions of a single particle for a long period of time, the force potential for this particle can be determined. The distribution should follow the Boltzmann distribution $\exp(-\phi(x, y)/kT)$, where $\phi(x, y)$ is the potential of the optical trap, k is the Boltzmann constant, and T is the absolute temperature. Therefore, this method is similar to the one that uses a quadrant detector.

The only difference in our method is the use of cameras that allow short exposure times to measure the equilibrium distribution of a particle in the trap. To determine the exposure time, the distance of travel of the particle during the exposure time, as well as the ratio of the distance versus the width of the potential, must be considered. The former must be sufficiently short to obtain reasonable accuracy in the position measurement. The latter is important because as this ratio decreases, the measured distribution is distorted such that more particles are distributed toward the center of the trap. This makes the measured potential sharper than the actual potential. Meanwhile, as the exposure time is decreased, it becomes more difficult to obtain images with sufficient quality to determine the coordinates of the particle with sufficient accuracy.

In general, we employ a 0.1 msec exposure time for particles in aqueous solutions and cell culture mediums. For this purpose, we use either a CCD camera equipped with a microchannel intensifier with electronic gating capability (ICCD-350F, Videoscope) or a high speed video (FASTCAM-ultima, PHOTRON, Japan).

V. Dragging Receptor Molecules along the Plasma Membrane and Finding Obstacles in the Path

Gold particles attached to membrane constituent molecules work not only as a probe, but also as a handle to move these molecules along the plasma membrane with the use of laser tweezers. In experiments with membrane proteins, latex particles also are used to label target proteins. By using the laser trap to apply restraining forces to particles, the experimenter is able to capture, move, and release particle–protein complexes in the membrane at will. For example, Kucik *et al.* (1991) and Schmidt *et al.* (1993) used laser tweezers to move particle–glycoprotein complexes to various locations on the cell surface and to restrain bead movement at specific locations in the cell. Gloushankova *et al.* (1995) used laser tweezers to place concanavalin A-coated beads at specific places on the surface of active lamellae.

A. Representative Dragging Experiments

Representative experiments in dragging receptor molecules are displayed in Fig. 5. In these experiments, latex particles of 210-nm ϕ (L210) coated with

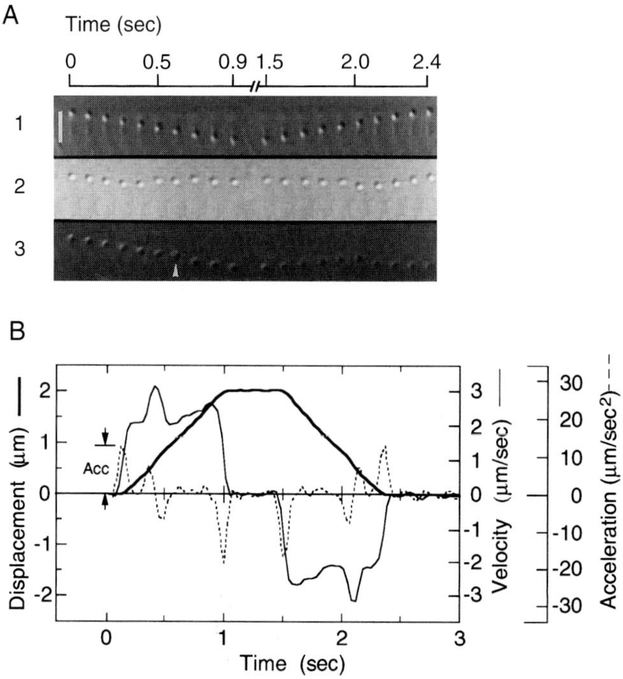

Fig. 5 Dragging of cell-surface TR labeled with latex beads (L210) using laser tweezers. The receptor-L210 complexes were dragged by moving the laser beam of the optical tweezers. The beam was moved 2 μm on the cell surface at a velocity of 2.4 μm/sec, and then was moved backward to the starting position (down and then up). The trapping laser power in these images was 100 mW, measured immediately after the objective lens, which provides a trapping force of 0.8 pN for L210. (A) Serial microscopic images of particles that were dragged by the laser tweezers. The images of the particles were obtained by video-enhanced differential interference contrast microscopy. Bar, 2 μm. (1) A particle that was dragged the entire distance (2 μm) and then back to the starting position. (2) A particle that escaped from the optical trap during the forward scan. The particle escaped ≈0.4 sec after starting the scan, and rapidly moved back toward the starting position (by 0.5 sec). At ≈2.0 sec, it was trapped again by the returning beam. (3) A particle that escaped during the backward scan of the laser trap. This particle was dragged to the end of the forward scan (2 μm away from the initial position) and escaped at ≈2.0 sec in the return portion of the trip. The particle showed retardation in the forward scan (arrowhead) near the position at which it escaped in the backward scan (from Sako and Kusami, 1995). (B) A typical dragging pattern of a particle shown as a function of time. Displacement (thick solid line), velocity (thin solid line), and acceleration (broken line) are shown as functions of time. These parameters were obtained every 33 msec (video frame rate) and were averaged over 3 successive points. Acc (initial acceleration) is defined as shown. The large variation in the velocity display at ≈0.4 sec (near the $\frac{1}{3}$ point) in the forward scan is probably not due to Brownian motion because a similar variation is seen at 2.1 sec (near the $\frac{2}{3}$ point) in the backward scan.

transferrin were bound to transferrin receptor (TR) molecules and captured by optical tweezers. The particle-TR complexes were moved laterally along the plasma membrane by scanning the trapping laser beam. The trapping force depends on the laser power and the particle properties: 0.8, 0.35, and 0.1 pN for L210 and 0.25, 0.1, and 0.05 pN for transferrin-coated 40-nm ϕ gold particles (G40) at incident laser powers of 100, 50, and 25 mW, respectively. [These powers were measured just after the beam exited from the objective lens of 100×, numerical aperture (NA) 1.3]. As long as the force exerted on the TR by the membrane or other cellular structures (such as the cytoskeleton–membrane skeleton and clathrin-coated structures) was less than the maximal force of the trap, the particles that were dragged along the plane of the membrane remained within the trap. The laser beam was moved a distance of 2 μm (forward and then backward, making a round trip) along the membrane plane in each direction at a velocity of 2.4 μm/sec. The particles visualized at the video rate are shown in Fig. 5A. In Fig. 5B, typical examples of displacement, velocity, and acceleration of a particle induced by the movement of the optical trap (the scan of the trapping laser beam) are shown as a function of the time after dragging was initiated. The large variation in the velocity at about one-third of the forward trip (at ≈0.4 sec) is probably not due to thermal fluctuation of the particle's movement because a similar variation is observed in the backward scan (2.1 sec). We speculate that these variations (around the peaks in the velocity display) reflect the passage of the TR over an obstacle or a barrier in the dragging path.

B. Typical Cases in Which the Trajectories of the Dragged Receptor Suggest the Presence of Barriers in the Dragging Path

The movements of the TR molecules as they are moved by the laser tweezers at different trapping forces are shown in Fig. 6. At a trapping force of 0.8 pN, the majority (67%) of the TR-L210 complexes that showed fast lateral diffusion ($D_{micro} > 1.5 \times 10^{-10}$ cm²/sec) were moved to the maximum extent of the laser scan (2 μm), then returned to the initial position (Fig. 6A, trajectory 1; the forward trajectories are shown in black, whereas the return trip is shown in green).

At lower trapping forces, the particles' trajectories under laser dragging became tottery. This tottering movement was not due solely to Brownian motion within the laser trap, because the trajectory of the return portion of the trip often agreed fairly well with that of the forward portion (Fig. 6A, trajectories 2 and 3). These results show that the trajectories are sensitive to the presence of obstacles in the dragging path.

With weaker trapping forces, the particles often escaped from the tweezers during either the forward or backward trip (Figs. 5B and 5C). When the particles escaped from the optical trap on the forward trip, they often rebounded very rapidly toward the initial starting position (toward the left in Fig. 6B, the red portions of the trajectories) for about 0.2 sec, then began to exhibit Brownian diffusion (the blue portions of the trajectories). When this fast rebound motion

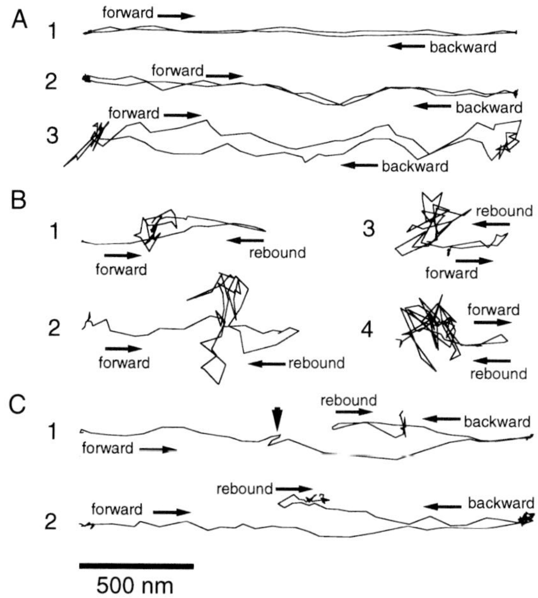

Fig. 6 Trajectories of TR-particle complexes, showing dragging by and escape from the laser trap. To produce these trajectories, the x and y coordinates of the particles were determined frame-by-frame, and then simply connected by straight lines. The accuracy of the coordinate determination has been shown to be better than 1.8 nm for particles fixed on a cover glass (Kusumi *et al.,* 1993). Arrows indicate the direction of the laser trap scan. In the presentation of the trajectories in this figure, the particles are initially moved from left to right (forward scan or outbound trip), then from right to left (backward scan or return trip). Bar, 500 nm. (A) Trajectories of TR-particle complexes dragged to the end of the scan. Under weaker trapping forces (2, 3), the particles showed tottering motions within the trap. This tottering motion was probably not due simply to Brownian motion, but rather to obstacles in the dragging path, because the trajectories in the backward scan coincide with those in the forward scan. (1) L210-TR dragged at a laser power of 100 mW (= trapping power of 0.8 pN). (2) L210-TR dragged at a laser power of 50 mW (=0.35 pN). (3) L210-TR dragged at a laser power of 25 mW (=0.1 pN). (B) Trajectories of particles that escaped during the forward scan. The particle-TR complexes were dragged from left to right. After the particles escaped from the optical trap, they often rebounded very rapidly toward the initial starting position and then began to exhibit Brownian diffusion. (1 and 2) L210 at laser powers of 100 mW (=0.8 pN, 1) and 50 mW (=0.35 pN, 2). (3 and 4) G40 at laser powers of 100 mW (=0.25 pN, 3) and 25 mW (=0.05 pN, 4). (C) Trajectories of particles that escaped during the backward scan. After the particles escaped from the optical trap, they often moved backward (in the direction of the turn point, for ≈0.2 sec) and began to exhibit Brownian diffusion. Near the position where the particle escaped, a deviation is often found in the forward trajectory (the vertical arrowhead in trajectory 1), which suggests that the particle hit the same boundary or obstacle in both directions. (1) L210 at 100 mW (=0.8 pN). (2) G40 at 25 mW (=0.05 pN).

is viewed in the video monitor, it looks as if the particle were catapulted out of the trap (like an arrow released from a bow), which suggests that the obstacle (or boundary or fence) is an elastic structure. This movement, however, cannot be the hurled motion of the particle off an elastic boundary. Because viscosity, rather than inertia, dominates the movements of submicrometer objects, the particles would stop as soon as they are released from the elastic boundary. Therefore, the fast rebound motion of the particles is likely to reflect the motion of the elastic boundaries restoring their original shapes. The effective elastic constant of the boundaries is in the range of 1–10 pN/μm, as described later (Fujiwara, Sako, Mellman, and Kusumi, unpublished observation; Sako and Kusumi, 1995).

Figure 6C shows the trajectories of particles that escaped during the backward scan (the green portions of the trajectories). After the particles escaped from the optical trap, they often moved backward (in the direction of the turn point, shown in red for $\approx$0.2 sec) and began to exhibit Brownian diffusion (shown in blue). In trajectory 1, near the position where the particle escaped on the return trip, a deviation is found in the forward trajectory (the vertical arrowhead in trajectory 1), which suggests that the particle hit the same boundary or obstacle in both directions.

Escape during the backward trip is of particular interest because it can readily be explained by the fence model (Fig. 2B) but not by the tether model (Fig. 2A). Approximately 6% of the particles dragged to the end of the forward trip escaped during the return trip. In addition, a deviation (small rebound) was often found in the forward trajectory near the position where the particle escaped in the backward trip (Fig. 6C, trajectory 1), which again suggests the presence of a barrier at the compartment boundaries (fence model rather than tether model).

VI. Laser Tweezers Together with SPT Can Reveal the Presence of Elastic Intercompartmental Barriers

A. Long-Time SPT Has Shown That the Plasma Membrane Is Compartmentalized with Regard to the Lateral Diffusion of Transmembrane Proteins and That Many Membrane Proteins Undergo Intercompartmental Hop Diffusion

Studies of SPT showed that the plasma membranes of a variety of mammalian cells, such as normal rat kidney fibroblastic (NRK) cells, Madin-Darby canine kidney (MDCK) epithelial cells, and mouse keratinocytes, are compartmentalized into many small domains of 0.1–1 μm^2 (0.4–1.2-μm diameter) with regard to the lateral diffusion of many membrane proteins, including TR, α_2-macroglobulin receptor, epidermal growth-factor receptor, and E-cadherin (Fig. 7, Kusumi *et al.*, 1993; Sako and Kusumi, 1994). Observation of the movements of TR and α_2-macroglobulin receptor in the plasma membrane of NRK cells for about 6 min showed that these molecules are confined within the compartments for

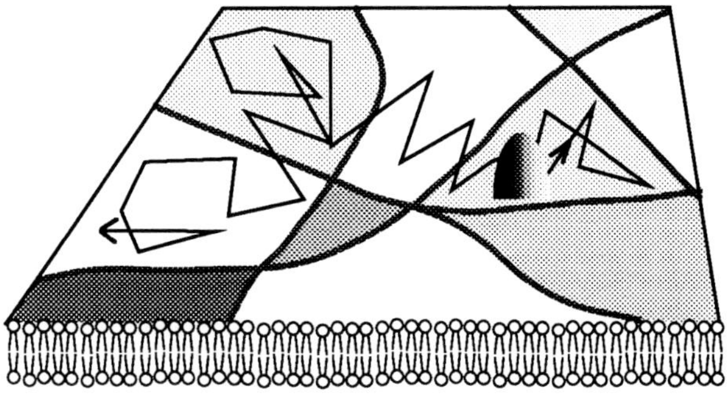

Fig. 7 A model showing the compartmentalized structure of the plasma membrane and the intercompartmental hop diffusion of membrane proteins. The plasma membrane is compartmentalized into many domains of 0.1–1.0 μm^2 for the diffusion of transmembrane proteins. In the case of transferrin receptor in NRK cells, the receptor molecules undergo almost free diffusion within a compartment (slowed only by the presence of other membrane proteins) to which they are confined for an average of ≈25 sec. The receptor molecules move from a compartment to one of the adjacent domains at a frequency of ≈0.04 sec^{-1} (2.4 intercompartmental hops/min) on average, and the long-range diffusion of receptors occurs as the result of successive intercompartmental movements. Therefore, the *macroscopic* diffusion rate is determined from the size of the compartment and the frequency of jumps between compartments, and gives the macroscopic D of $\approx 2.4 \times 10^{-11}$ cm^2/sec.

25 sec on average, then hop to an adjacent compartment. Within a domain, these receptor molecules undergo rapid lateral diffusion that is indicative of free diffusion (microscopic diffusion coefficient $\approx 10^{-9}$ cm^2/sec), whereas the rate of long-range diffusion of these receptors is smaller by a factor of 30. The macroscopic diffusion rate is determined by the size of the compartment *and* the frequency of intercompartmental hops, which is irrelevant to the microscopic diffusion rate. Partial destruction of the microfilaments or microtubules dramatically changed the motional modes of these receptors (Sako and Kusumi, 1994).

B. Barrier-Free Path Length (BFP)

Edidin *et al.* (1991, 1994) were first to carry out experiments to obtain the barrier-free path length. They moved transmembrane-(H-$2D^b$) and glycosylphosphatidyl–inositol (GPI)-anchored (Qa2) major histocompatibility complex class I molecules in the plasma membrane. The protein–gold complexes were dragged by laser tweezers until they escaped from the trap when they encountered barriers. Thus, the barrier-free path lengths (BFP) could be obtained. They were 3.5 and 8.5 μm for the transmembrane and GPI-anchored species, respectively, at 34°C. The BFP of a mutant H-$2D^b$ molecule, which possesses only 4 amino acid residues in the cytoplasmic domain (vs. 31 residues in the wild type), was greater than that of the wild type by a factor about two (Edidin *et al.,* 1994). These results suggest that the cytoplasmic domain of the transmembrane protein is involved in the regulation of the lateral movements of these proteins.

At lower trapping forces, compartments of comparable sizes as observed by SPT could be detected (Sako and Kusumi, 1995). At a maximal dragging force of 0.25 to 0.8 pN, the TR molecules undergoing confined diffusion with a high microscopic diffusion coefficient (D_{micro}) of $\approx 10^{-9}$ cm^2/sec (this applies to 90% of the TR in NRK cells) could be dragged past the intercompartmental boundaries in their path. At lower dragging forces, between 0.05 and 0.1 pN, TR tended to escape from the laser trap at the boundaries, suggesting that the force acting on TR from the boundaries was greater than these dragging forces.

The important characteristics of these escapes are that (a) they occurred in *both* the forward and backward directions of dragging (which cannot be explained by tethering of TR to the cytoskeleton); (b) the boundaries are elastic; and (c) the BFP for each molecule on the average was half of the confinement size estimated by SPT observation, which further indicates that TR escaped at the compartment boundaries. (Because the average start point of dragging is the center of the compartment, the dragging distance is expected to be half of the compartment size, if TR always escapes at the boundaries.)

C. Examining the Membrane–Skeleton Fence Model

Because variation of the particle size (40- and 210-nm particles that are on the extracellular surface of the plasma membrane) hardly affects the diffusion rate and behavior in the dragging experiments, and because treatment with cytochalasin D or vinblastin affects the movements of TR, the boundaries are likely to be present in the cytoplasmic domain (Sako and Kusumi, 1994, 1995). The rebound of the particle-TR complexes when they escape from the laser tweezers at the compartment boundaries suggests that the boundaries are elastic structures. These results are consistent with the proposal that the compartment boundaries consist of a membrane-associated part of the cytoskeleton (membrane–skeleton fence model, Fig. 8).

In this model, the membrane skeleton provides a barrier to the free diffusion of membrane proteins due to steric hindrance (the space between the membrane and the cytoskeleton is too small to allow the cytoplasmic portion of the membrane protein to pass), and thus compartmentalizes the membrane into many small domains of 0.1–1 μm^2 (Fig. 7; Boal and Boey, 1995; Jacobson *et al.,* 1995; Sheetz *et al.,* 1980; Sheets *et al.,* 1995). The membrane proteins can escape from one domain and move to adjacent compartments due to the dynamic properties of the membrane skeleton: The distance between the membrane and the skeleton may fluctuate over time, thus giving the membrane proteins an opportunity to pass over the mesh barrier; the membrane skeleton may dissociate from the membrane; and the membrane–skeleton network may form and break continuously due to the dissociation–association equilibrium of the cytoskeleton. In

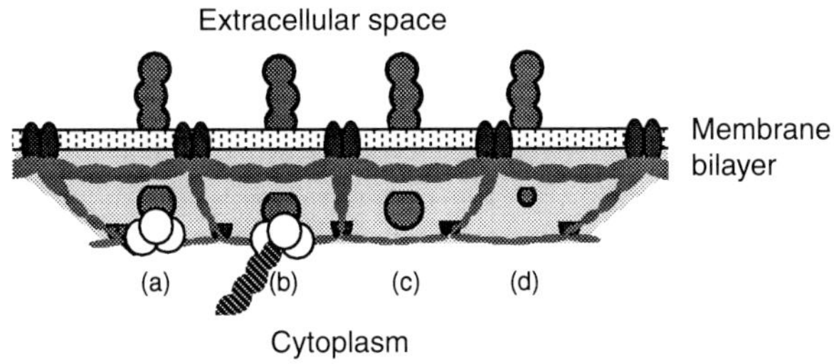

Fig. 8 Membrane–skeleton fence model. In this figure, the plasma membrane is viewed from inside a cell. The bilayer portion of the membrane, the transmembrane proteins (light shading) and those connecting the membrane skeleton to the membrane (darker shading), the cytoplasmic proteins bound to membrane proteins (open circles, in a and b), and the binding of this complex to the cytoskeleton (b) are shown. The membrane skeleton is in close proximity to the cytoplasmic surface of the plasma membrane. The cytoplasmic domain of the membrane protein (or its complex with a cytoplasmic protein) collides with the membrane skeleton and cannot readily move to an adjacent compartment (a, c, d). If the cytoplasmic domain is smaller (d), the protein moves to an adjacent compartment more readily. Binding of membrane proteins to the membrane skeleton (b, but direct binding of transmembrane proteins to the membrane skeleton also occurs) has been found for a variety of proteins. Binding can be detected by using laser tweezers to drag the membrane protein. In many cases, the membrane skeleton itself is undergoing macroscopic diffusion as well as oscillative motion without any real displacement. Some membrane skeletons show directed, active transport-type movements, and the transmembrane proteins bound to such skeletons also undergo similar directed-type movements.

addition, membrane protein molecules that possess sufficient kinetic energy will be able to pass over the boundaries.

Confined lateral diffusion and intercompartmental hop diffusion of membrane proteins have been observed for a variety of membrane proteins in all cells studied thus far. We propose that compartmentalization of the plasma membrane by a membrane–skeleton/cytoskeleton meshwork (membrane–skeleton fence structure) is a basic feature of the plasma membrane. For individual protein species, more specific mechanisms such as direct binding to the cytoskeleton may be at work. However, what should be emphasized here is that the general fence effect of the membrane skeleton is superimposed on the specific effect that applies to individual protein species. In the case of E-cadherin, some molecules that are bound to the flexible cytoskeleton (possibly thin actin filaments) "feel" the presence of the membrane skeleton fence as they move about with the attached cytoskeleton (Sako and Kusumi, unpublished observation).

VII. Studying the Fence and Tether Effects of the Membrane Skeleton Using Laser Tweezers

A. Binding and Transport of Membrane Proteins by the Membrane Skeleton

Binding of membrane proteins to the membrane skeleton has been found for almost all proteins investigated thus far, including the receptors for transferrin,

EGF, and α2-macroglobulin, E- and T-cadherins, and band 3 anion channel in erythrocytes (Fig. 8b; Discher *et al.,* 1994; Hikichi, 1995; Kusumi *et al.,* 1993; Sako and Kusumi, 1994; Tomishige, 1995). These bound proteins undergo various types of motion. Some show no motion, whereas others show oscillative movements without real translation. Some proteins show long-range translational diffusion while they are apparently bound to the membrane skeleton (which is known because these particles cannot be dragged more than 100 nm by laser tweezers). Some show directed transport-type movements probably due to active movements of the cytoskeleton to which they are bound (Kusumi *et al.,* 1993; Fujiwara, Sako, Mellman, and Kusumi, unpublished observations).

B. How to Play a Tug of War with the Cellular Membrane Skeleton

Laser tweezers provide a useful method for distinguishing receptor molecules bound to the membrane skeleton from those unbound, but corralled, by the membrane skeleton. Immunoglobulin G (IgG) Fc receptor II-B2 (FcγR) molecules, expressed in CHO cells (Miettinen *et al.,* 1992) and labeled with gold particles coated with a small number of anti-FcγR Fab, were dragged by laser tweezers. The results of the dragging experiments are displayed in Fig. 9, in which the particle position is plotted against the time after dragging. The position of the center of the optical trap is also indicated at each time point (every 33 msec). Because the optical trap was moved at a constant rate, the latter falls on a straight line.

Two representative cases were found for the initial response of FcγR after dragging: In one case, the FcγR molecule starts to lag behind the optical trap immediately after dragging is initiated (Fig. 9A), and in the other case, the FcγR molecule initially followed the trap for a while (up to 0.9 sec or to $d_f = 0.5\ \mu$m) and then started to lag behind the trap (Fig. 9B). The former can be explained by the tether model, whereas the latter can be explained by the fence model in which the FcγR molecule followed the trap until it hit a compartment boundary and then started to lag behind the trap because the force exerted by the membrane skeleton began to act on the FcγR molecule.

In the experiments shown in Fig. 9, laser tweezers are used to play a tug of war with the cell's membrane skeleton, as shown in Fig. 10. The FcγR molecule-gold complex (shown as a particle) is pulled by both laser tweezers (spring on the right) and the membrane skeleton (spring on the left). Because the force applied by the laser tweezers is always balanced by that exerted by the membrane skeleton, the following equation is obtained (also see Fig. 9B).

$$k_l\,(r_l - r_c) = k_c\,r_c, \tag{1}$$

where k_l is the elastic constant of the laser tweezers (as described in the previous chapters, the force potential in the optical trap can be described by a harmonic potential when the particles are located near the center of the trap, l represents laser tweezers), r_l is the displacement of the center of the optical trap after the particle starts to feel the force from the membrane skeleton (Figs. 9B and 10),

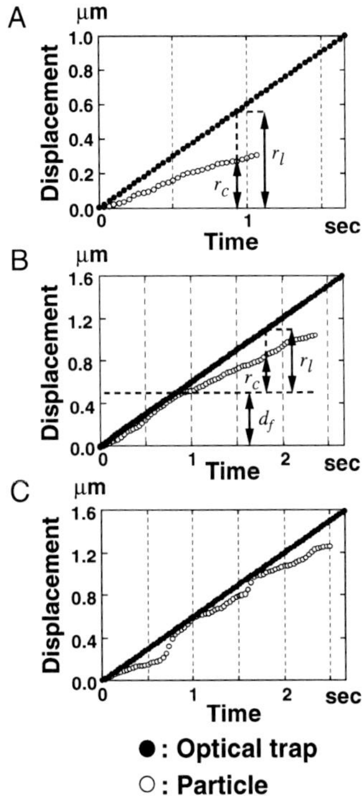

Fig. 9 Both fence and tether effects of the membrane skeleton can be studied and distinguished by using laser tweezers. FcγR molecules, expressed in CHO cells and labeled with gold particles coated with a small number of anti FcγR Fab, were dragged by laser tweezers. The particle position is plotted against the time after dragging (open circles). The position of the center of the optical trap is also indicated at each time point (every 33 msec, closed circles). Because the optical trap was moved at a constant rate. Two representative cases of the initial response of FcγR after dragging are shown in (A) and (B). (A) An FcγR molecule started to lag behind the optical trap right after dragging was initiated, which can be explained by the tether model (Fig. 2A). (B) An FcγR molecule initially followed the trap for a while, then started to lag behind the trap at ≈0.5 μm (d_f) from the start position (0.9 sec), which can be explained by the fence model; that is, the FcγR molecule followed the trap until it hit a compartment boundary, then began to lag because the force exerted by the membrane skeleton acted on the FcγR molecule. (C) A receptor molecule lagged behind the optical tweezers (0.1–0.6 μm from the start) but caught up later (0.1–0.6 sec or 0.1–0.2 μm from the start), suggesting that the receptor hit a boundary at 0.1 μm from the start point but then crossed the boundary at 0.7–0.8 sec or 0.2–0.4 μm from the start. The same receptor appeared to hit a second boundary near 0.6 μm from the start point (1 sec) and to cross it again (0.8–1.0 μm, 1.6–1.7 sec), but when the receptor hit the third boundary, it could not pass the boundary (the boundary won the tug of war with the laser tweezers) and escaped from the optical trap.

r_c is the displacement of the particle after it starts to feel the force from the membrane skeleton (i.e., the extension of the membrane skeleton; Figs. 9B and 10; c represents the cytoskeleton), and k_c represents the elastic constant of the

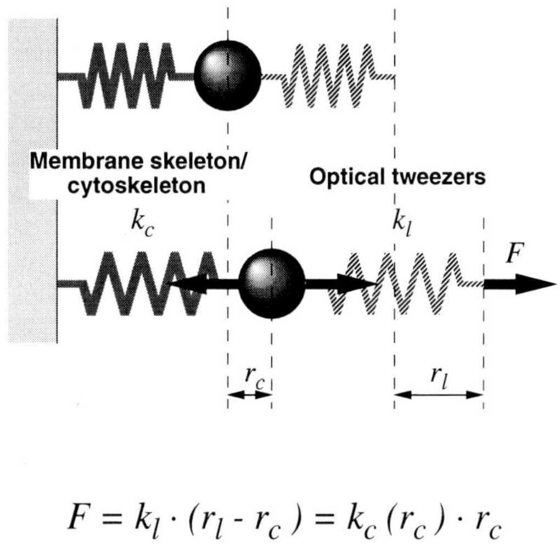

Fig. 10 A tug of war between the cellular membrane skeleton and the laser tweezers. The FcγR molecule-gold complex (shown as a particle) is pulled by both the laser tweezers (spring on the right) and the membrane skeleton (spring on the left). The elasticity of the membrane skeleton, k_c, can be calculated from each point after the membrane skeleton starts to pull the receptor in the display in Fig. 9 because r_l and r_c are determined from the plot and k_l is calibrated. Variable k_c is generally in the range of 1–10 pN/μm, which is a very weak spring.

membrane skeleton. To include the higher order effect of the displacement on k_c, it can be treated as a function of r_c.

The elasticity of the membrane skeleton, k_c, can be calculated from each point after the membrane skeleton starts to pull on the receptor in the display in Fig. 9, because r_l and r_c are determined from the plot and k_l is calibrated; k_c is generally in the range of 1–10 pN/μm, which is a very weak spring. If a penny were to be hung on such a spring (of course, on the earth's surface), the spring would stretch for several miles. However, even such a weak spring would be sufficiently strong to regulate the movements of membrane proteins in the plasma membrane.

In the experiment shown in Fig. 9C, the receptor lagged behind the optical tweezers (0.1–0.6 sec or 0.1–0.2 μm from the start) but caught up later (0.7–0.8 sec or 0.2–0.4 μm from the start), suggesting that the receptor hit a boundary but then crossed it as the receptor was dragged more strongly by the laser tweezers (as the particle moves away from the center of the trap, greater force acts on the particle). The same receptor appeared to hit the second boundary near 0.6 μm from the start point (1 sec) and to cross it again (0.8–1.0 μm, 1.6–1.7 sec), but when the receptor hit the third boundary, it could not pass the boundary and escaped from the optical trap (the boundary won the tug of war with the laser tweezers).

As exemplified here, both the fence and tether effects of the membrane skeleton can be studied by laser tweezers. The use of laser tweezers can provide various types of important information about the nature of the interactions between membrane proteins and the membrane skeleton, the structure of the membrane skeleton, and its mechanical properties.

Nevertheless, the way cells control the corralling and binding effects of the membrane skeleton has yet to be elucidated. It is likely that cells are using the fence effect and active transport by the membrane–skeleton/cytoskeleton to assemble specific membrane proteins into specialized domains. However, exactly how cells do this has not been discovered and remains one of the most important issues in membrane biology.

VIII. SPT and Laser Tweezers Are Opening Up Possibilities for Scientists to Learn about the Molecular Mechanics in Cells by Directly Handling Single Molecules in Living Cells

SPT and laser tweezers have opened new ways to study the processes by which cells organize various molecules in the plasma membrane. Whereas cells use "molecular hands" (molecular interactions) to organize molecules with nanometer precision by forces on the piconewton level, we can now use "optical hands" to manipulate single molecules with similar spatial precision and with forces of a comparable magnitude; hence we can *interrupt or imitate the molecular hands of cells.* This will help to elucidate the mechanisms and principles behind the formation of supramolecular complexes in the plasma membrane.

The membrane skeleton plays a pivotal role in the molecular organization of the plasma membrane. The fence and binding effects on the movements and assembly of membrane proteins are key processes for cell surface organization. SPT and laser tweezers, which can deal with single (or a few) molecules at nanometer/piconewton precision in living cells, are especially suitable for studying the interactions between membrane proteins and the membrane–skeleton/cytoskeleton.

Acknowledgment

We thank Dr. Paul Wiseman for his critical reading of this manuscript.

References

Anderson, C. M., Georgiou, G. N., Morrison, I. E. G., Stevenson, G. V. W., and Cherry, R. J. (1992). Tracking of cell surface receptors by fluorescence digital imaging microscopy using a charge-coupled device camera. Low-density lipoprotein and influenza virus receptor mobility at 4°C. *J. Cell Sci.* **101,** 415–425.

Baudhuin, P., van der Smissen, P., Beauvois, S., and Courtoy, P. J. (1989). Molecular interactions between colloidal gold, proteins, and living cells. *In* "Colloidal Gold: Principles, Methods, and Applications, Vol. 2" (M. A. Hayat, ed.), pp. 1–17. New York: Academic Press.

Bennett, V. (1990). Spectrin-based membrane skeleton: A multipotential adaptor between plasma membrane and cytoplasm. *Physiol. Rev.* **70,** 1029–1065.

Bennett, V., and Gilligan, D. M. (1993). The spectrin-based membrane skeleton and micron-scale organization of the plasma membrane. *Annu. Rev. Cell Biol.* **9,** 27–66.

Boal, D. H., and Boey, S. K. (1995). Barrier-free paths of directed protein motion in the erythrocyte plasma membrane. *Biophys. J.* **69,** 372–379.

Cherry, R. J. (1992). Keeping track of cell surface receptors. *Trends Cell Biol.* **2,** 242–244.

de Mey, J. R. (1983). Colloidal gold probes in immunocytochemistry. *In* "Immunocytochemistry (Practical Applications in Pathology and Biology)" (J. M. Polak and S. van Noorden, eds.), pp. 83–112. Bristol, UK: Wright PSG.

Discher, D. E., Mohandas, N., and Evans, E. A. (1994). Molecular maps of red cell deformation: Hidden elasticity and *in situ* connectivity. *Science* **266,** 1032–1035.

Edidin, M. (1992). Patches, posts and fences: Proteins and plasma membrane domains. *Trends Cell Biol.* **2,** 376–380.

Edidin, M. (1993). Patches and fences: probing for plasma membrane domains. *J. Cell Sci. Suppl.* **17,** 165–169.

Edidin, M., Kuo, S. C., and Sheetz, M. P. (1991). Lateral movements of membrane glycoproteins restricted by dynamic cytoplasmic barriers. *Science* **254,** 1379–1382.

Edidin, M., Zúñiga, M. C., and Sheetz, M. P. (1994). Truncation mutants define and locate cytoplasmic barriers to lateral mobility of membrane glycoproteins. *Proc. Natl. Acad. Sci. USA* **91,** 3378–3382.

Geerts, H., De Brabander, M., and Nuydens, R. (1991). Nanovid microscopy. *Nature* **351,** 765–766.

Gelles, J., Schnapp, B. J., and Sheetz, M. P. (1988). Tracking kinesin-driven movements with nanometre-scale precision. *Nature* **331,** 450–453.

Ghislain, L. P., and Webb, W. W. (1993). Scanning-force microscope based on an optical trap. *Optics Lett.* **18,** 1678–1680.

Ghosh, R. N., and Webb, W. W. (1994). Automated detection and tracking of individual and clustered cell surface low density lipoprotein receptor molecules. *Biophys. J.* **66,** 1301–1318.

Gloushankova, N. A., Krendel, M. F., Sirotkin, V. A., Bonder, E. M., Feder, H. H., Vasiliev, J. M., and Gelfand, I. M. (1995). Dynamics of active lamellae in cultured epithelial cells: Effects of expression of exogenous N-*ras* oncogene. *Proc. Natl. Acad. Sci. USA* **92,** 5322–5325.

Hikichi, Y. (1995). A study on the regulation mechanism of protein movements on the cell surface by single particle tracking and laser tweezers: comparison of T-cadherin, N-cadherin, and transferring receptor. Tokyo: University of Tokyo. Ph. D. thesis.

Hitt, A. L., and Luna, E. J. (1994). Membrane interactions with the actin cytoskeleton. *Curr. Opin. Cell Biol.* **6,** 120–130.

Jacobson, K., Sheets, E. D., and Simson, R. (1995). Revisiting the fluid mosaic model of membranes. *Science* **268,** 1441–1442.

Kucik, D. F., Kuo, S. C., Elson, E. L., Sheetz, M. P. (1991). Preferential attachment of membrane glycoproteins to the cytoskeleton at the leading edge of lamella. *J. Cell Biol.* **114,** 1029–1036.

Kusumi, A., and Sako, Y. (1996). Cell surface organization by the membrane skeleton. *Curr. Opin. Cell Biol.* **8,** 566–574.

Kusumi, A., Sako, Y., and Yamamoto, M. (1993). Confined lateral diffusion of membrane receptors as studied by single particle tracking (nanovid microscopy). Effects of calcium-induced differentiation in cultured epithelial cells. *Biophys. J.* **65,** 2021–2040.

Leunissen, J. L. M., and de Mey, J. R. (1989). Preparation of gold probes. *In* "Immuno-Gold Labeling in Cell Biology" (A. J. Verkleij and J. L. M. Leunissen, eds.), Boca Raton, FL: CRC Press, pp. 3–16.

Luna, E. J., and Hitt, A. L. (1992). Cytoskeleton-plasma membrane interactions. *Science* **258,** 955–964.

Mecham, R. P., Whitehouse, L., Hay, M., Hinek, A., and Sheetz, M. P. (1991). Ligand affinity of the 67-kD elastin/laminin binding protein is modulated by the protein's lectin domain: Visualization of elastin/laminin-receptor complexes with gold-tagged ligands. *J. Cell Biol.* **113,** 187–194.

Miettinen, H. M., Matter, K., Hunziker, W., Rose, J. K., and Mellman, I. (1992). *J. Cell Biol.* **116,** 875–888.

Nelson, W. J. (1992). Regulation of surface polarity from bacteria to mammals. *Science* **258,** 948–955.

Qian, H., Sheetz, M. P., and Elson, E. (1991). Single particle tracking (analysis of diffusion and flow in two-dimensional systems). *Biophys. J.* **60,** 910–921.

Roth, J. (1983). The colloidal gold marker system for light and electron microscopic cytochemistry. *In* "Immunocytochemistry 2" (J. M. Polak, and S. van Noorden, eds.), pp. 217–284. Bristol, UK: Wright PSG.

Sako, Y., and Kusumi, A. (1994). Compartmentalized structure of the plasma membrane for lateral diffusion of receptors as revealed by nanometer-level motion analysis. *J. Cell Biol.* **125,** 1251–1264.

Sako, Y., and Kusumi, A. (1995). Barriers for lateral diffusion of transferrin receptor in the plasma membrane as characterized by receptor dragging by laser tweezers: Fence versus tether. *J. Cell Biol.* **129,** 1559–1574.

Saxton, M. J. (1993). Lateral diffusion in an archipelago. Single-particle diffusion. *Biophys. J.* **64,** 1766–1780.

Saxton, M. J. (1994). Anomalous diffusion due to obstacles: a Monte Carlo study. *Biophys. J.* **94,** 683–688.

Saxton, M. J. (1995). Single-particle tracking: Effects of corrals. *Biophys. J.* **69,** 389–398.

Saxton, M. J. (1996). single-particle tracking: New methods of data analysis. *Biophys. J.* **70,** A334.

Schmidt, C. E., Horwitz, A. F., Lauffenburger, D. A., and Sheetz, M. P. (1993). Integrin cytoskeletal interactions in migrating fibroblasts are dynamic, asymmetric, and regulated. *J. Cell Biol.* **123,** 977–991.

Sheets, E. D., Simson, R., and Jacobson, K. (1995). New insights into membrane dynamics from the analysis of cell surface interactions by physical methods. *Curr. Opin. Cell Biol.* **7,** 707–714.

Sheetz, M. P., Schindler, M., and Koppel, D. E. (1980). Lateral mobility of integral membrane proteins is increased in spherocytic erythrocytes. *Nature* **285,** 510–512.

Sheetz, M. P., Turney, S., Qian, H., and Elson, E. L. (1989). Nanometre-level analysis demonstrates that lipid flow does not drive membrane glycoprotein movements. *Nature* **340,** 284–288.

Simson, R., Sheets, E. D., and Jacobson, K. (1995). Detection of temporary lateral confinement of membrane proteins using single particle tracking analysis. *Biophys. J.* **69,** 989–993.

Tomishige, M. (1995). Analysis of the membrane skeleton network of human erythrocytes by single particle tracking and manipulation: Elastic deformability of the cell and regulation of the movements of membrane proteins. Tokyo: University of Tokyo. M. S. thesis.

Wang, Y.-L., and Silverman, J. D., Cao, L.-G. (1994). Single particle tracking of surface receptor movement during cell division. *J. Cell Biol.* **127,** 963–971.

Yoshitake, M., Imagawa, S., Ishikawa, E., Niitsu, Y., Urushizaki, I., Nishiura, M., Kanawawa, R., Kurosaki, H., Tachibana, S., Nakazawa, N., and Ogawa, H. (1982). Mild and efficient coagulation of rabbit Fab′ and horseradish peroxidase using a maleimide compound and its use for enzyme immunoassay. *J. Biochem.* **92,** 1413–1422.

CHAPTER 11

In Vivo Manipulation of Internal Cell Organelles

Harald Felgner,[*] Franz Grolig,[†] Otto Müller,[*] Manfred Schliwa[*]

[*] Adolf-Butenandt-Institut, Zellbiologie
University of Munich
80336 Munich, Germany

[†] Institut für Allgemeine Botanik und Pflauzenphysiologie
University of Giessen
35390 Giessen, Germany

I. Introduction

Since Ashkin and Dziedzic (1987) first used an optical trap to manipulate biological objects (viruses and bacteria), single-beam laser tweezers have frequently been employed to hold, move, and deform cells and subcellular particles. Examples include, but are not restricted to, duplication of yeast cells (Ashkin *et al.,* 1987), manipulation of nuclei and organelles in plant cells (Ashkin and Dziedzic, 1989; Leitz *et al.,* 1994) and protozoa (Aufderheide *et al.,* 1992), displacement of chromosomes or chromosome fragments in cultured cells (Berns *et al.,* 1989; Seeger *et al.,* 1991), blockage of axonal transport (Martenson *et al.,* 1993), and cell sorting (Buican, 1991; for reviews, see Block, 1990; Kuo and

Copyright © 1998 by Academic Press. All rights of reproduction in any form reserved.
0091-679X/98 $25.00

Sheetz, 1992; Weber and Greulich, 1992). Several of these studies have demonstrated convincingly that intracellular organelles up to the size of nuclei can be displaced or deformed without damaging effects. Optical tweezers-based techniques therefore offer novel opportunities for subcellular manipulation and the study of cell processes that require precise positioning of cell organelles.

Here we describe some basic techniques for the intracellular manipulation of cell organelles using the giant freshwater amoeba, Reticulomyxa, and the green alga, Spirogyra, as model systems. Both cell models afford damage-free manipulation of mitochondria, particles, or membrane tubes, thus providing insights into the forces required to move organelles.

II. Optical Tweezers Setup

The optical tweezers used are formed by a continuous-wave neodymium: yttrium aluminum garnet Nd : YAG laser, wavelength 1064 nm (Spectron, Rugby, England), expanded by a telescope and coupled into an upright light microscope (Zeiss Axioskop with Plan-NEOFLUAR 100×/1.30, Oberkochen, Germany), as shown in Fig. 1. The laser beam is fixed relative to the image plane, but objects near the coverglass may be manipulated in all three directions by moving the microscope stage via stepping motors (Märzhäuser EK32 and MCL, Wetzlar, Germany) and computer control (Apple Macintosh IIfx, Cupertino, U.S.). Cells were monitored by phase-contrast optics (4× magnification), and images were recorded with a CCD-camera (Hamamatsu C2400/C3077, Herrsching, Germany) and a video recorder (Panasonic AG-7350, Düsseldorf, Germany). The experi-

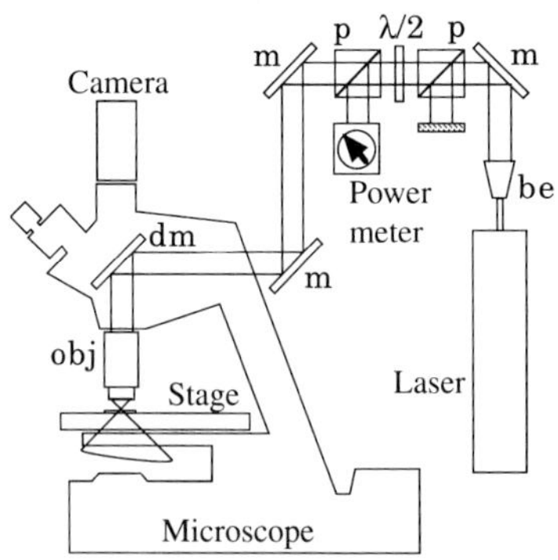

Fig. 1 Schematic diagram of the optical tweezers setup. be, beam expander; m, mirrors; p, polarizing beam-splitter cubes; λ/2, λ/2-wave plate; dm, dichroic mirror; obj, objective. The dichroic mirror is used to reflect the infrared laser beam into the microscope objective and to transmit the image to the CCD camera. The camera is connected to an image-processing board, video recorder, and monitor. The motorized stage is driven via computer control.

ments on cell process formation in Reticulomyxa and on motor forces in Spirogyra were performed with this instrument. The determination of motor forces in Reticulomyxa were done with a similar setup described elsewhere (Ashkin and Dziedzic, 1989; Ashkin *et al.,* 1987).

III. Reticulomyxa: Organelle Movement and Artificial Membrane Tubes

A. Cell Preparation

The giant amoeba Reticulomyxa possesses a central cell body (10–50 μm in diameter) and a peripheral feeding network of fine cytoplasmic strands that extend several centimeters from the cell body (Euteneuer *et al.,* 1989). The reticulate peripheral network displays rapid movements of organelles at velocities up to 25 μm/sec. Reticulomyxa is a useful model system for the study of the forces underlying organelle transport and the formation of artificial membrane tubes *in vivo* with the use of optical tweezers.

Reticulomyxa was grown in petri dishes in spring water and was fed wheat germ (Koonce *et al.,* 1986). For light microscopy, pieces of the cell body were sucked into a micropipette and placed on coverslips in culture medium. After 1 to 3 hr they developed a new reticulopodial network. The cell body was removed, and the coverslip with the remaining network was placed on a slide. The fine strands with a diameter down to 50 nm can easily be seen in phase-contrast microscopy.

B. Organelle Movement

To measure the forces driving single mitochondria along microtubules within fine strands of the reticulopodial network, a mitochondrium simply has to be trapped with a force greater than the molecular motor force (Ashkin *et al.,* 1990). As the laser power is slowly decreased, the mitochondrium should escape at the moment when the light force equals the motor force. From the calibrated light forces and the number of motors involved in the movement of the mitochondrium, the force of a single molecular motor can be determined *in vivo.*

In practice, the experiment was made more difficult by the heavy traffic of mitochondria and other organelles in a typical strand. Whereas the mitochondria are clearly distinguishable from other particles by their size and optical density, the problem lies in the fast accumulation of organelles in the trap during the slow reduction of the laser power. Therefore, the procedure was to trap a moving mitochondrium in a narrow strand with a diameter of 0.3 μm or less at a high light force and to quickly reduce the laser power to a preset value. In narrow strands there are fewer moving particles, and the probability of disturbing organelle accumulations in the trap is reduced. The membrane of such fine strands enclosing a bundle of one to six microtubules bulges as the mitochondria with a greater diameter (320 ± 70 nm) than the strands pass along the microtubules. Due to the small diameter of the trapping region, in broader strands of the network, particles may elude the laser trap as they pass by even at high laser powers. To determine the molecular forces, the laser power at which the mitochondrium was trapped and the power to which it was reduced immediately to let the particle escape again were averaged.

The laser force on a particle depends both on the relative refractive indices of the particle and its surrounding medium and on the particle size and geometry. Assuming a uniformity concerning these parameters in the population of mitochondria in the cell, it is possible to calibrate the light forces on mitochondria in a cell homogenate. The mitochondrium is trapped and accelerated relative to the buffer of known viscosity. From the velocity at the point the mitochondrium escapes from the trap, the force can be calculated by Stokes' Law. The velocity of trapped particles before and after trapping may be recorded and compared. The lack of a significant change in speed before and after manipulation with the laser is an indication that the motors are not damaged by the laser light.

The molecular motors in Reticulomyxa are thought to be cytoplasmic dynein (Euteneuer *et al.*, 1988; Koonce *et al.*, 1986). As shown by electron microscopy, there appear to be one to four motors moving a single mitochondrium (Ashkin *et al.*, 1990). Together with the measured force to hold a mitochondrium in the trap, this leads to a force of approximately 2.6 piconewtons (pN) per dynein molecule (Ashkin *et al.*, 1990).

C. Artificial Membrane Tubes

The strands of the reticulopodial network of Reticulomyxa consist of microtubule bundles enclosed by the plasma membrane of the cell. By trapping a mitochondrium or another organelle at a high laser power in a fine strand, and by moving the microscope in a direction approximately perpendicular to the strand, we were able to form artificial membrane tubes. Instead of using an organelle to draw out the membrane, it is possible to incubate the cells on the coverslips with micrometer-size latex beads, letting them internalize the beads, which then may serve as defined handles.

We were able to form membrane tubes of lengths up to several micrometers. If pulled out too fast or too long, the organelle, together with the membrane tube, slipped out of the laser trap. The newly formed tube did not contain any cytoskeletal network when pulled out fast, as suggested by the limp appearance and speedy retraction and disappearance of the membrane tube within seconds. Furthermore, no traffic of organelles could be observed in the newly formed membrane tube. Holding a membrane tube stretched out for more than a few seconds enabled the formation of new cytoskeletal elements inside the membrane and allowed for organelle traffic to be established. When the membrane tube was released again, it nevertheless retracted and disappeared into the initial strand, but the qualitatively different, stiff appearance of the membrane tube over the period of retraction was clearly visible. It is unclear whether the formation of the new cytoskeleton is due to polymerization of new microtubules within or to transport of existing microtubules into the membrane tube. By bringing a membrane tube into contact with another nearby strand, we were able to fuse the membranes of tube and strand. The artificial strand formed in this way, if stabilized by cytoskeletal elements, persisted for long periods of time and behaved in a manner indistinguishable from that of natural strands. Figure 2 depicts a sequence of new membrane tube formation and its fusion with a neighboring strand.

IV. Spirogyra: Cytoplasmic Streaming

In most plant cells an extensive directional bulk movement of cytoplasmic particles termed *cytoplasmic streaming* occurs along dynamic and extensive tracks of actin, with myosin(s) being the putative motor(s) (reviewed by Kamiya, 1981; Williamson, 1993). The large and particularly translucent, filamentous green alga *Spirogyra crassa* (Zygnematales and Charophyceae) permits the detailed observation of cytoskeletal activity and manipulations *in vivo* by video-enhanced microscopy (Grolig, 1990; Grolig, 1992; Sawitzky and Grolig, 1995).

Cells of a length/width ratio greater than or equal to 3 were prepared for optical trapping of mitochondria by removal of the optically interfering, spiral chloroplast bands from the peripheral cytoplasm by gentle centrifugation (250 g for 10 min) in a swing-out rotor. For centrifugation, filaments were mounted with their long axis parallel to the centrifugal force between the halves of a bloc of solidified agar (3% w/v); (Fujii *et al.*, 1978). Dislocation of the chloroplasts is not harmful to the cell, which is capable of restoring its intracellular arrangement of organelles within 3 to 5 days (Kuroda and Kamiya, 1991; Wisselingh, 1909).

After centrifugation, numerous small vesicles (diameter 1 μm), some membrane tubules and many mitochondria (diameter 1 μm, length 3–5 μm) remain in the thin (5 μm) peripheral cytoplasm along dynamic tracks of F-actin. Due to the large cell diameter (160 μm), slight compression of the cell cylinder, and narrow depth of the peripheral cytoplasm, a virtually two-dimensional field of observation is obtained for observing and recording the movements of these organelles. All these organelles appear dark in phase contrast, are easily discernible, and seem to use identical tracks, as judged either from different particles traveling in file on the same track or from membrane tubule coalignment with such a track. During movement, the mitochondria usually are aligned with their long axis parallel to the underlying actin track. A constant stream of moving, small vesicles helps in visualizing the unidirectional, dynamic tracks so that the optical trap can be positioned with sufficient precision in front of a more slowly (less steadily) advancing mitochondrium.

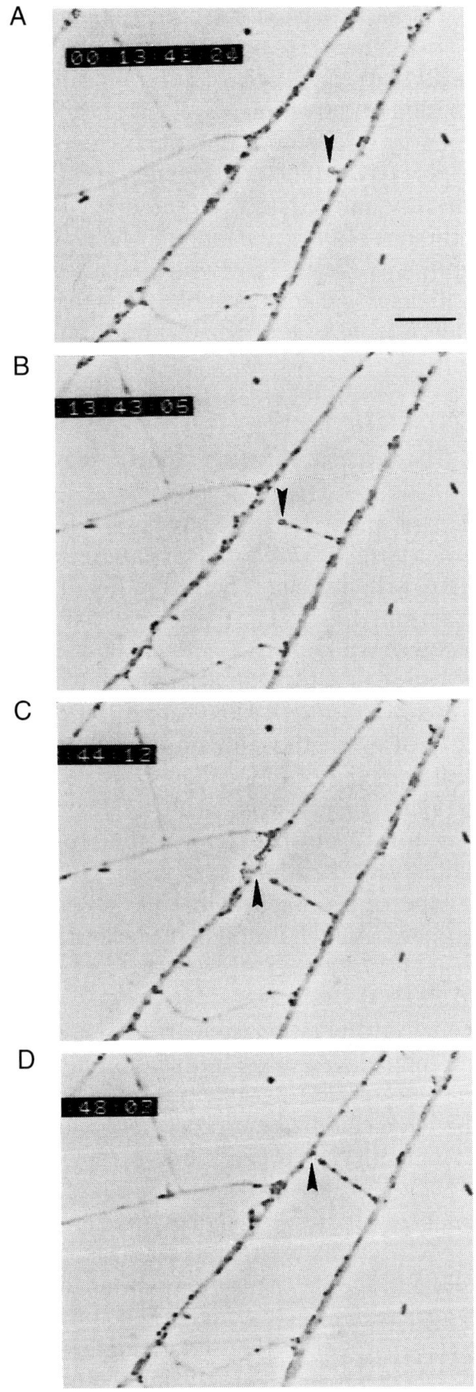

Fig. 2 Creating an artificial cell extension. By grabbing a particle in one strand of a reticulomyxa network (A), a new cell process (arrowhead) can be pulled out (B). Upon contact with a neighboring strand (C), the two membranes fuse instantaneously, forming a stable connection that slides along the two parallel strands (D). The entire sequence took place in four sec. Bar, 5 μm.

Preliminary force calibrations of the trap against Stoke's drag by acceleration of trapped, isolated mitochondria in media of known viscosity indicate that the minimum force needed for efficient trapping of a mitochondrium *in vivo* is in the range of 10 pN. With increased force of the trap, the mitochondria often aligned themselves with their long axis parallel to the optical axis of the trap (i.e., about normal according to the axis of the F-actin track). If the mitochondrium was not set free immediately after trapping, numerous small vesicles traveling on the same track readily accumulated in the trap. The trapped mitochondrium usually escaped upon accumulation of these vesicles if the trapping force had been set toward the minimum force needed for trapping. This behavior may be explained by a change of optical parameters in the trap upon vesicle accumulation, diminishing the force acting on the trapped mitochondrium. The sketches of Fig. 3A and

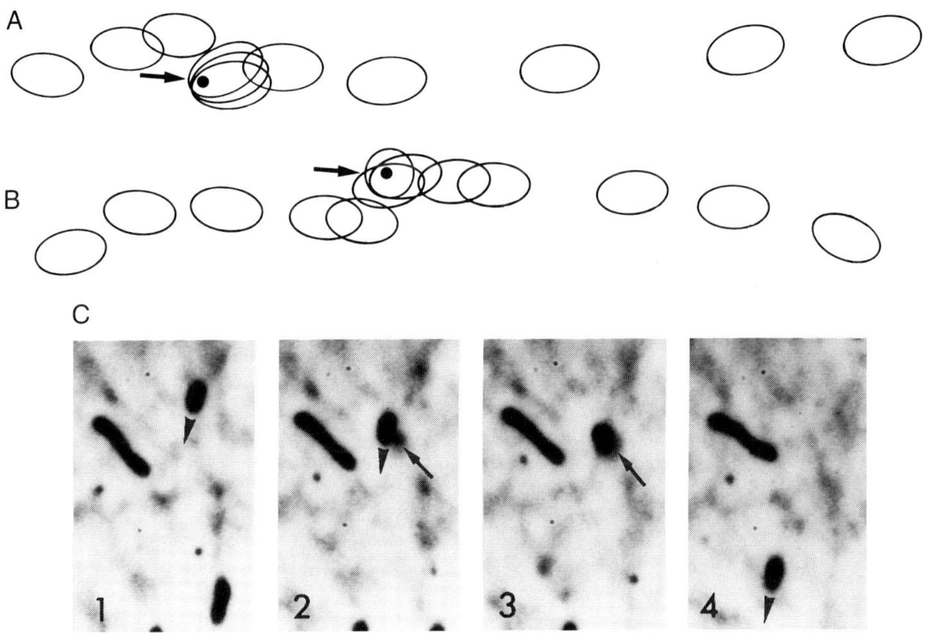

Fig. 3 (A) and (B) Sketches of two mitochondrial translocation sequences with transient trapping of the mitochondria (ellipsoids). Both mitochondria move from left to right. Arrows point to the stationary center of the trap (black dot). The mitochondrium of track (B) aligned its long axis parallel to the axis of the optical tweezers while trapped (circle, long axis perpendicular to direction of translocation). (C) Sequence of micrographs (1 to 4), showing details of the translocation sequence in (B). The trap, marked by arrows in (2) and (3), is active and has accumulated small vesicles (grey cloud) traveling on the same track as the mitochondrium which is trapped in (3). Arrowheads indicate direction of movement of the mitochondrium in the other micrographs. The video image was processed (average and digital contrast enhancement) with a real-time video processor (MulticonR, Leica, Bensheim, Germany) and photographed with a freeze-frame recorder (RGB-version, Polaroid Corp., Cambridge, MA, U.S.) equipped with a 35-mm camera adaptor.

escape after having been trapped. Both mitochondria show a slight lateral displacement when being trapped, but readily escape and move on as if this lateral displacement did not affect the contact to the F-actin track. The mitochondrium shown in Fig. 3B transiently orients its long axis normal to the track. This rearrangement apparently did not affect its capability to escape and move on. Excerpts of the video recording of this mitochondrium are shown in Fig. 3C.

Acknowledgment

This work was supported by the Deutsche Forschungsgemeinschaft (SFB266).

References

Ashkin A., and Dziedzic, J. M. (1987). Optical trapping and manipulation of viruses and bacteria. *Science* **235,** 1517–1520.

Ashkin, A., and Dziedzic, J. M. (1989). Internal cell manipulation using infrared laser traps. *Proc. Natl. Acad. Sci. USA* **86,** 7914–7918.

Ashkin, A., Dziedzic, J. M., and Yamane, T. (1987). Optical trapping and manipulation of single cells using infrared laser beams. *Nature* **330,** 769–771.

Ashkin, A., Schütze, K., Dziedzic, J. M., Euteneuer, U., and Schliwa, M. (1990). Force generation of organelle transport measured *in vivo* by an infrared laser trap. *Nature* **348,** 346–348.

Aufderheide, K. J., Du, Q., and Fry, E. S. (1992). Directed positioning of nuclei in living Paramecium tetraurelia: Use of the laser optical force trap for developmental biology. *Dev. Genet.* **13,** 234–240.

Berns, M. W., Wright, W. H., Tromberg, B. J., Profeta, G. A., Andrews, J. J., and Walter, R. J. (1989). Use of a laser-induced optical force trap to study chromosome movement on the mitotic spindle. *Proc. Natl. Acad. Sci. USA* **86,** 4539–4543.

Block, S. M. (1990). Optical tweezers: A new tool for biophysics. *In* "Non-invasive Techniques in Cell Biology" (J. K. Foskett and S. Grinstein, ed.), pp. 375–402. New York: John Wiley & Sons.

Buican, T. N. (1991). Automated cell separation techniques based on optical trapping. *Am. Chem. Soc. Symp. Ser.* **464,** 59–72.

Euteneuer, U., Koonce, M. P., Pfister, K. K., and Schliwa, M. (1988). An ATPase with properties expected for the organelle motor of the giant amoeba Reticulomyxa. *Nature* **332,** 176–178.

Euteneuer, U., Johnson, K. B., Koonce, M. P., McDonald, K. L., Tong, J., and Schliwa, M. (1989). *In vitro* analysis of cytoplasmatic organelle transport. *In* "Cell Movement" (F. D. Warner and J. R. McIntosh, ed.), pp. 155–167. New York: Alan R. Liss.

Fujii, S., Shimmen, T., and Tazawa, M. (1978). Light-induced changes in membrane potential in Spirogyra. *Plant Cell Physiol.* **19,** 573–590.

Grolig, F. (1990). Actin-based organelle movements in interphase Spirogyra. *Protoplasma* **155,** 29–42.

Grolig, F. (1992). The cytoskeleton of the zygnemataceae. *In* "The Cytoskeleton of the Algae" (D. Menzel, ed.), pp. 165–193. Boca Raton: CRC Press.

Kamiya, N. (1981). Physical and chemical basis of cytoplasmic streaming. *Annu. Rev. Plant Physiol.* **32,** 205–236.

Koonce, M. P., Euteneuer, U., McDonald, K. L., Menzel, D., and Schliwa, M. (1986). Cytoskeletal architecture and motility in a giant freshwater amoeba, Reticulomyxa. *Cell Motil. Cytoskel.* **6,** 521–533.

Kuo, S. C., and Sheetz, M. P. (1992). Optical tweezers in cell biology. *Trends Cell Biol.* **2,** 116–118.

Kuroda, K., and Kamiya, N. (1991). Cytoplasmic movement in Spirogyra during and after centrifugation. *Proc. Jpn. Acad.* **67,** 78–82.

Leitz, G., Weber, G., Seeger, S., and Greulich, K. O. (1994). The laser microbeam trap as an optical tool for living cells. *Physiol. Chem. Phys. Med. NMR* **26,** 69–88.

Martenson, C., Stone, K., Reedy, M., and Sheetz, M. P. (1993). Fast axonal transport is required for growth cone advance. *Nature* **366,** 66–69.
Sawitzky, H., and Grolig, F. (1995). Phragmoplast of the green alga Sporogyra is functionally distinct from the higher plant phragmoplast. *J. Cell Biol.* **130,** 1359–1371.
Seeger, S., Manojembashi, S., Hutter, K.-J., Futerman, G., Wolfrum, J., and Greulich, K. O. (1991). Application of laser optical tweezers in immunology and molecular genetics. *Cytometry* **12,** 497–504.
Weber, G., and Greulich, K. O. (1992). Manipulation of cells, organelles, and genomes by laser microbeam and optical trap. *Int. Rev. Cytol.* **133,** 1–41.
Williamson, R. E. (1993). Organelle movements. *Annu. Rev. Plant Physiol. Plant Mol. Biol.* **44,** 181–202.
Wisselingh, C. V. (1909). Zur Physiologie der Spirogyrazelle. *Beihefte zum Botanischen Centralblatt* **12,** 133–210.

CHAPTER 12

Optical Chopsticks: Digital Synthesis of Multiple Optical Traps

Justin E. Molloy
Department of Biology
University of York
York, YO1 5DD, United Kingdom

I. Introduction

The theory and design of optical traps is discussed in Chapter 1 of this volume. This chapter describes how multiple optical traps may be synthesised by rapid scanning of a single laser beam. For many optical trapping applications it is advantageous to be able to produce two or more traps simultaneously. One technique is to split the trapping laser beam into two separate light paths using a polarising beam splitter (Simmons *et al.,* 1996) and then to recombine the light using a second polarising beam splitter. This allows two, independently controllable optical traps to be produced. The method presented here (first described by Visscher *et al.,* 1993) allows two or more traps to be synthesised by rapidly scanning a single laser beam, using fast beam deflectors, to generate

METHODS IN CELL BIOLOGY, VOL. 55
Copyright © 1998 by Academic Press. All rights of reproduction in any form reserved.
0091-679X/98 $25.00

multiple traps that are not truly simultaneous but which function by time sharing the laser power between several positions. The principle of operation is similar to that of television, in which a picture is created using a single electron gun that is scanned much faster than the persistence time of the phosphors used on the screen. Multiple optical traps can be created by rapid scanning of a single laser beam because the viscosity of the solution (usually water) is sufficiently high to provide positional persistence. For the simplest case, persistence time depends only on passive diffusion of the object being trapped; in other applications external forces acting on the object must also be considered.

The system described here was designed for measuring single-molecule mechanical interactions between actin and myosin. The electronic circuit was optimized to produce the two optical traps used for these experiments. However, it should be moderately straightfoward to modify the circuit to produce four or more traps if required. For a scanning system to be of any benefit it must be under computer control, because this allows quantitative measurements to be made while the optical traps are being manipulated. For this reason the author has given an outline of a custom-built IBM PC interface.

II. Trapping Configurations

A. Condition 1: Simplest Case: Holding an Object

Most optical traps work well only to a radius of about 300 nm. During the time when the trap is "off" servicing another position (t_{off}), the object may diffuse completely out of the trapping region. This must be avoided! For a spherical object, the diffusion coefficient, $D = kT/6\pi\eta a$ where k is the Boltzmann constant, T is the temperature, η is the viscosity, and a is the radius of object (e.g., for a 1-μm latex bead, $D = 4.2*10^{-13}$ m^2sec^{-1}). The mean square motion in time t; $<x^2> = 2Dt$ and $x_{rms} = (2t_{off}\cdot (kT/6\pi\eta a))^{0.5}$; irrespective of the laser power used, the maximum t_{off} is $\approx$100 ms. The "on" period required to restore the bead close to its resting position will depend on the stiffness of the optical trap, κ. The time constant τ for a spherical bead $= 6\pi\eta a/\kappa$ (from Stokes' law; $F = 6\pi\eta av = \kappa x$ and $\tau = x/v$) (Fig. 1).

B. Condition 2: Trap Used to Manipulate an Object

For many applications optical traps are used to manipulate objects in solution. If the microscope stage is driven by hand, stage velocities of about 200 $\mu msec^{-1}$ will be generated. This means that the object will be displaced from the trap center position by an amount $x = 6\pi\eta av/\kappa$ during t_{on} (e.g., if $\kappa = 0.05$ pN/nm, $a = 0.5$ μm, stage velocity $= 200$ $\mu msec^{-1}$; $x \approx 40$ nm). The maximum t_{off} is now very short because during t_{off} the object will be dragged out of the trapping radius at the same speed as the solution flow (maximum $t_{off} = (300 - 40$ nm)/

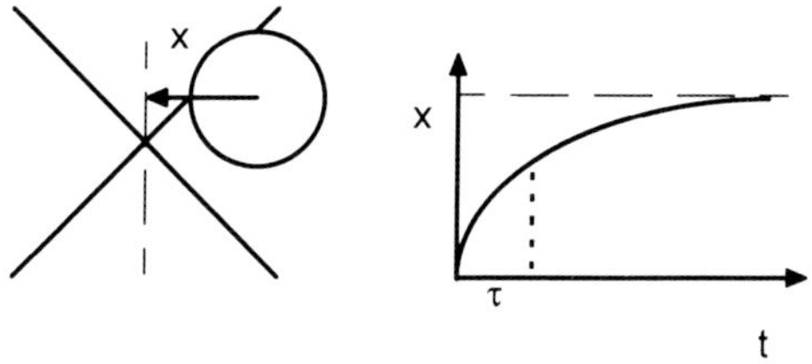

Fig. 1 When using multiple traps synthesized from a single laser beam, objects must be "serviced" by the trap frequently enough that they do not diffuse out of the effective trapping radius (≈300 nm) during the "off" interval. During the "on" interval, the object will be returned toward the trap center with an exponential time course. At commonly used trapping stiffnesses (<0.5 pN/nm) Brownian motion makes the observed motion much more noisy than the schematic time course.

$200\mu msec^{-1}$ = 1.3 msec). The chopping frequency must therefore be at least 500 Hz for a two-trap system and proportionately faster if more traps are required. Lower frequencies are acceptable if the stage is moved slowly (e.g., if great care is taken when using the stage controls). However, scanning frequencies much lower than 500 Hz will lead to the objects "jumping out" of the trap as they are manipulated in solution.

C. Condition 3: Trap Used to Exert a Static Force

If the optical traps are used to exert a static force on a trapped object, then the object will be pulled away from its resting position during t_{off} and restored toward the resting position during t_{on} (Fig. 2). If a high frequency symmetrical duty cycle is used (i.e., two traps with equal dwell times being synthesized) the motion will be nearly triangular.

$$A \approx x_0\kappa.\ t_{off}/6\pi\eta a$$

where A equals the amplitude of peak-to-peak motion, and x_0 equals the offset from the trap rest position (κx_0 = force).

For quantitative applications, high positional stability is required, and the back-and-forth motion that can occur when the trap exerts a static force can interfere with the measurements being made. If very rapid chopping of the laser beam position is accomplished, this chopping-induced motion can be minimized.

For the bead-actin-bead dumbbell, most often used to study acto-myosin interactions (Finer *et al.,* 1994) (Fig. 2), the effective radius is that of both beads. Also, because the bead pair is held close to the coverslip surface, the viscous drag will be approximately doubled (Faxen's law; Svoboda and Block, 1994b). Thus, for the two-bead configuration, the observed motion is about four times smaller than calculated from the preceding equation. We find that by chopping the trap positions at 10 kHz the resulting motion is actually ≈1 nm root mean square (2.5 nm peak to peak) when 2 pN of tension is applied to the actin filament.

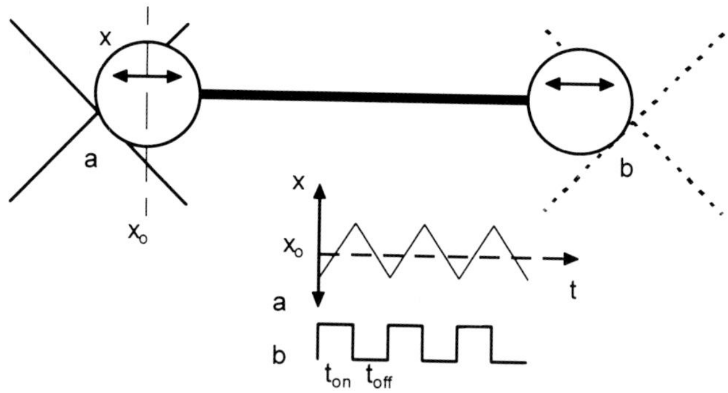

Fig. 2 Diagram to show the experimental arrangement used during studies of single molecule acto-myosin interactions. Two plastic beads (≈1-μm diameter) are bound at either end of a single-actin filament, and these are held in two synthesized optical traps; by moving the traps apart, the actin filament can be held under tension. At high chopping frequencies the force produced by trap (a) is counterbalanced by the viscous force exerted on the beads as they are dragged through the solution. The beads move a small distance left and then right as the trap is chopped from position (a) to (b) (graph, inset). The distance moved depends on the tension exerted, the viscous drag, and the chopping frequency (see text).

III. Mechanical Deflectors (Mirrors)

It is possible to deflect a mirror either with galvanometer or piezoelectric drivers at frequencies up to approximately 10 kHz. Because mirrors are relatively heavy, it is difficult to achieve higher resonant frequencies than this, and the motions produced will be underdamped unless electronic feedback is applied. The fastest rise time for a feedback-controlled mechanical deflector (i.e., with good damping) is probably about 200 μsec. The best that the author has achieved for a two-axis piezo-driven system is 1 msec. Allowing four optical traps to be produced, but trap stability and effective stiffness was sufficient only for crude manipulation of objects (condition 1 earlier) and insufficient for making quantitative measurements. At reasonable chopping frequencies (>100 Hz), the rise time (transit time from one trap position to another) of such a device is a significant fraction of the total duty cycle, and there is loss of effective laser trapping power because of this "dead time" (Fig. 3). Also, chopping-induced motion of the two-trap dumbbell configuration (Fig. 2) is unacceptably high (>50 nm for 2 pN tension).

IV. Acousto-optic Deflectors

Acousto-optic deflectors (AODs) are solid-state devices that have no moving parts. They consist of a crystal of transparent material in which a traveling

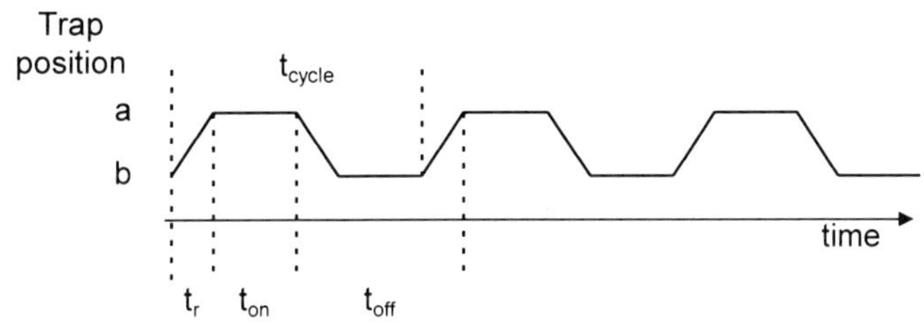

Fig. 3 At the high scanning frequencies required to produce low chopping-induced motion under load, the slow rise time, t_r, of mechanical deflector systems makes them impractical because the scanning frequency is limited to about $(4tr)^{-1}$ = 250 Hz, so the chopping-induced motion will be large. Also, the transit time between trap locations results in excessive "dead time" and reduction in trap stiffness.

acoustic wave is generated by a piezo transducer bonded to one of its surfaces. The acoustic wave generates a refractive index wave that behaves as a sinusoidal grating. When tilted to the Bragg angle most of the light that enters the crystal is reflected into the first-order diffraction beam. The frequency of the acoustic wave determines the spacing of the grating and hence the angle of the first-order beam. To give x and y axis motion two AOD scanners must be mounted orthogonally to each other. AODs exhibit near ideal properties of low drift and noise, no creep, high speed, and good resolution. Use of AODs is vital for synthesis of multiple traps of high stability. They are the only devices known that can be driven fast enough to keep chopping-induced motion under load to below 5 nm. The most commonly used material for diffracting near-infrared laser light (we use 1064 nm) is tellurium oxide (chosen for its high "figure of merit," which translates into high diffraction efficiency: ≈80% of the light being reflected into the first order). The acoustic velocity in this material is ≈700 m sec^{-1}, and the transit time or propagation delay across the incident laser beam as the frequency is modulated to produce a new diffraction spacing is complete in ≈1μsec (laser beam diameter/acoustic velocity). The acoustic waves generated by the piezo transducer are radio frequency (30–80 MHz), and commercially available driver electronics have either digital or analogue input. The computer interface described here should work for either type of driving electronics (just remove the D/A converters for the digital synthesized driver and use the digital signal directly).

Our optical trap is based around a Zeiss Axiovert microscope (Molloy *et al.*, 1995b); the optical path is shown in Fig. 4. We have experimented with two different AOD devices: (*a*) NEOS (NEOS Technologies, Melbourne, Florida, N45035-3-6.5deg-1.06 scanner, N72006xy Bragg mount and N64010-100-2ASDFS digital synthesizer driver and (*b*) Isle Optics (Taunton, Somerset, UK): TS100 mounted scanners and SD100-4A voltage controlled RF driver.

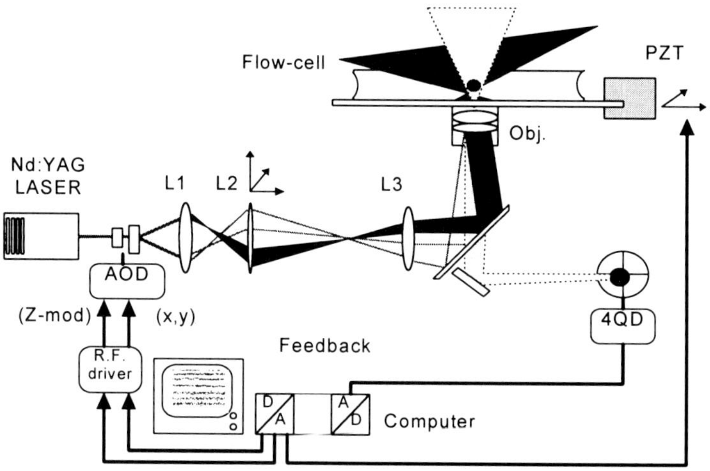

Fig. 4 The AOD's are driven by a high-power RF driver that has x, y, and z modulation inputs. The control signals are provided by a computer interface, either by direct digital input (NEOS synthesizer diver) or by analogue voltages (Isle Optics voltage-controlled oscillators). To synthesize multiple optical traps with a single-input laser beam, the x, y signals are chopped between different sets of coordinates. The voltage or digital signals are chopped at high frequency (>10 kHz), and the rise time of the analogue control signal should be better than 1 μsec. Motion of the output laser beam from the AODs is collimated by a short focal length lens (L1; f = +40 mm). This lens in combination with a second, longer focal length lens (L2; f = +150 mm) acts as a beam expander. The beam then enters the Axiovert microscope via the fluorescence port, passes through a third lens (L3; f = +100 mm) and enters the back aperture of the microscope objective (Acroplan, 100× 1.2 NA). A halogen lamp is used to produce a bright-field image of the trapped object (usually a latex microsphere), which is cast on a 4-quadrant photodetector (4-Q-D). This gives a signal proportional to the position of the object with nanometer precision and 10 kHz bandwidth. An IBM PC 486 66-MHz computer is used to collect data (via A/D converters), to control the AOD devices and microscope piezo substage (PZT) using a custom-built interface board (Fig. 5). The apparatus is described in more detail elsewhere (Veigel *et al.*, submitted).

V. Position: Control and Noise

If digitally synthesised optical traps are being used to make quantitative measurements of force or movement, sources of positional noise must be minimized. Here we assume that the pointing stability of the laser is very good and that the mechanical stability of the rest of the apparatus (microscope stage and other optical components) is also good. However, it is usually the mechanical stability of the objective lens and microscope stage that are the weakest links. Stability of the entire system must be thoroughly checked. However, there are several sources of noise that arise directly from synthesizing multiple traps by the chopping method. One should always bear in mind not only the amplitude, but also the bandwidth (or frequency) of any noise source.

- *Chopping-induced motion:* Chopping-induced motion under load (see earlier) can be minimized by using a high chopping frequency. We use a chopping frequency of 10 kHz, which keeps this source of noise to about 1 nm root mean square.
- *Bit noise:* Bit noise arises from steps in position associated with the change in signal for each digital bit. For most quantitative measurements the apparatus will be used in two modes: *free running,* in which the trap positions are held fixed and movement of the object being trapped is monitored and *feedback,* in which the position of the object is held fixed by driving the trapping laser position so as to compensate for any external forces applied to the object. The consequences of bit noise under these operating conditions are slightly different:

  Free-running mode: With trap positions held fixed, bit noise is not particularly important, but it should be remembered that each digital bit will usually have about $\frac{1}{4}$ bit of analogue noise. After conversion this noise will contaminate the position signal. Also, if each digital bit represents a large motion, then control of the trap position may be too coarse. In our system, the compromise chosen was to make 1 digital bit produce 2 nm of movement with a maximum range of motion ($\approx$13 useful bits) $\approx$8 μm.

  Feedback mode: One or maybe more trap positions are driven by either a digital or analogue feedback servo so as to maintain the object position constant. We use a digital servo-loop to control the position of one of the latex beads. The machine code used to perform the servo operation is given in Fig. 5. The position is accurately determined by a four-quadrant position detector (Molloy *et al.,* 1995b) and the signal used to calculate proportional and velocity errors. These terms are added to the starting position and also to an arbitrary forcing function. The resulting value is used to servo the laser position by sending a suitably scaled signal to the AOD driver. Bit noise translates into steps in the force signal. One digital bit of positional noise when multiplied by the trap stiffness gives the bit noise in terms of force (e.g., If trap stiffness = 0.05 pN nm^{-1} and one bit produces 2-nm trap movement, then bit noise = 0.1pN/bit.
- *Other sources of system noise:* Most of the positional noise in AODs arises from the control signal. If the requirement is for two traps to be positioned up to 10 μm apart and for both traps to be stable to better than 1nm, then the control signal must be good to 1 part in 10,000. The advantage of using digital control is that 16-bit data (1:65536) can be transferred across a noisy laboratory fairly easily. Handling an analogue signal with the same noise level is much less straightforward.
- *Brownian motion:* The effects of Brownian motion have been discussed elsewhere (Molloy and White, 1997). The important point is that this motion should be sampled over its full bandwidth. The effects of Brownian motion on the system being studied may be quite variable, but for most systems it may be unwise to average the signal away by filtering.

```
; The Assembler Feedback Loop:
; from FedBak32.asm

; Main loop
; Do the x-axis

A1: PUSH   CX
    MOV    DX,STAT1          ; Poll the A/D Done bit
    MOV    BX,OLDXERROR:  16 bit number
A2: IN     AL,DX
    TEST   AL,10000000B      ; Wait until done bit set
    JZ     A2

    MOV    DX,AD0DATA        ; get x A/D value
    IN     AX,DX             ; clear done bit
    STOSW                    ; save and Inc DI,Inc DI

    SUB    AX,INITQ4DX       ; get error from centre position (dx)

    MOV    DX,AX             ; Get forcing function from lookup
table
    LODSW
    ADD    AX,DX
    MOV    OLDXERROR,AX: save the error for next pass
    SUB    BX,AX             ; put change in error (dx/dt) in BX

    CWDE                     ; Sign extend AX register to EAX
    IMUL   EAX,GAIN          ; 32 Bit signed mul by proportional
gain
    MOV    ECX,EAX           ; 32 Bit temporary save in ECX

    MOVSX  EAX,BX
    IMUL   EAX,GAINVEL       ; signed multiply by velocity gain

    ADD    EAX,ECX           ; add proportional and velocity signals
    ADD    EAX,OLDAODX       ; subtract combined error signal
32Bits
    ROR    EAX,16            ; divide by 65536
    MOV    DX,AODXa
    OUT    DX,AX             ; Output the new AOD position
    ADD    DI,8192*4-2       ; Save the value in data array
    STOSW

    ROL    EAX,16            ; Switch low and high words convert
                             ; back to 32 bit
    MOV    OLDAODX,EAX
    SUB    DI,8192*4         ; Back to where we were

; Now do the y axis

    MOV    BX,OLDYERROR

    MOV    DX,AD1DATA

    IN     AX,DX
    STOSW                    ; Save and Inc DI,Inc DI

    SUB    AX,INITQ4DY       ; get error  from centre position (dy)
                             ; AX holds error
    PUSH   AX

    LODSW                    ; lookup forcing function for "b" trap
    ADD    AX,INITAODXb      ; add to start posn.
    MOV    DX,AODXb          ; move the "b" trap !
    OUT    DX,AX

    POP    AX

    MOV    OLDYERROR,AX
    SUB    BX,AX             ; get change in error (dy/dt)
    CWDE                     ; Sign convert AX to EAX

    IMUL   EAX,GAIN          ; signed multiply by proportional gain

    MOV    ECX,EAX           ; temporary save in ECX
    MOVSX  EAX,BX            ; get dy/dt
    IMUL   EAX,GAINVEL       ; signed multiply by velocity gain
    ADD    EAX,ECX           ; add proportional and velocity signals

    NEG    EAX               ; Switch sign for the 'y' axis only
    ADD    EAX,OLDAODY       ; subtract the combined error signal

    ROR    EAX,16            ; Move msb 16 bits down to AX

    MOV    DX,AODYa
    OUT    DX,AX             ; output the new AOD value
    ADD    DI,8192*4-2       ; save the value in data array
    STOSW

    ROL    EAX,16            ; Back to 32 bits

    MOV    OLDAODY,EAX

    SUB    DI,8192*4         ; Back to where we were

    POP    CX
    DEC    CX                ; Too far  to LOOP !
    JZ     A3                ; use JMP instruction
    JMP    A1                ; Do it CX times

A3: JMP    FFINI
```

Fig. 5 Digital servo of trap position using 386DX (or better) machine code. For speed, the CPU registers are used where possible, and integer arithmetic is used throughout. Rounding errors are limited by using 32-bit registers and dividing the result by 65,536 (reducing the result to a 16-bit value) at the end of the calculation. AODXa, AODxb, AODya are addresses of the multiplexed, AOD output registers. OLDAODX, OLDAODY hold the previous positions; INITQ4DX, INITQ4DY hold the starting positions measured by the four-quadrant detector; AD0DATA and AD1DATA address the A/D converters for the four-quadrant signals; OLDXERROR and OLDYERROR are the previous error values (used to calculate the differential error).

VI. Computer Control of Trap Positions

The best device for manual control of trap and microscope stage positions is undoubtedly the computer mouse. How this is implemented is computer-language dependent, but for C and QBASIC programming, the easiest way to communicate with the IBM PC mouse is via DOS interrupt 33 Hex. Among other things, there are calls that will read in the mouse x,y movement and the "button press" status (Duncan, 1986). Using different "button press" combinations a three-button mouse can be used to control up to eight optical traps together with the microscope stage (no button pressed). If only two traps are required, left and right buttons can be used to control each trap independently, and by pressing both buttons the motion of both traps can be locked together. By using double clicks the traps can be turned on and off and so forth.

Data collection in free-run mode is a fairly straightforward computer programming exercise. However, feedback mode will usually require machine code; Fig. 5 is a listing of the core part of the machine code used as a servo for the trap position. The basis of the servo is to calculate a proportional and differential error signal, then to combine that with an arbitrary forcing function contained in a look-up table, and finally to add this to the initial start position of the object. By using a personal computer (IBM 486, 66 MHz or better), the calculations can be completed in <20 μsec, which allows data collection and servo control for two channels (x and y) at a 25-kHz sample rate per channel. There are many commercial laboratory interface cards that offer 12-bit A/Ds and D/As operating at 100 kHz or more throughput. However, few of these boards offer simultaneous A/D acquisition on multiple channels, and none provide the necessary circuitry for hardware chopping between output registers. We built a simple interface board that contains several A/D and D/A converters with a common start conversion signal so that data acquisition and output is simultaneous for all channels. Trap positions can be chopped at high speed using a software loop (in response to a hardware interrupt). However, we found that there was a significant overhead in terms of computing time. Instead, we made a hardware chopping circuit that uses an electronic signal to switch the D/A converter data input lines between two sets of storage registers. Figure 6 shows the layout of the custom-built computer interface board. There are many companies that offer prototyping boards with the bus interface laid out as a printed circuit, and these allow the board to be made fairly readily.

VII. Other Developments

High efficiency of the AOD output depends on light entering the scanners at, or close to, the Bragg angle. As the driving frequency, and hence the spacing, is changed so this input angle should be adjusted. This is not possible, so the deflectors' output is flat only over a rather narrow range. The flat output range can be extended by use of the Z-modulation input to correct for fall off at the edges or other regions where efficiency falls or rises. This signal can be derived from a look-up table (the address of which is taken directly from the digital x,y value used to control the position). At present, we are building a hardware look-up table to perform this operation at high speed.

Interface board:

- **IBM PC/AT prototype board Interface**
  - 3* 74LS244
  - 2* 74LS245
  - 1* 74LS138
  - 1* 74LS139
  - 1* 74LS00
  - 1* 74LS10
- **Additional address decoding:**
  - 2* 74LS138
- **A/D converters (12 bit, 100kHz)**
  - 4* Burr-Brown ADS7804P
  - 1* 74LS08
  - 1* 74LS173
- **Programmable oscillator (600kHz-0.005Hz); A/D Start conversion.**
  - 1* SPG - 8640AN
  - 1* 74LS374
  - 1* 74LS123
- **Multiplexed Output registers**
  - 8* 74LS374
  - 1* NE555P
- **D/A converters (16 bit, 1μs):**
  - 4* Burr-Brown DAC712P

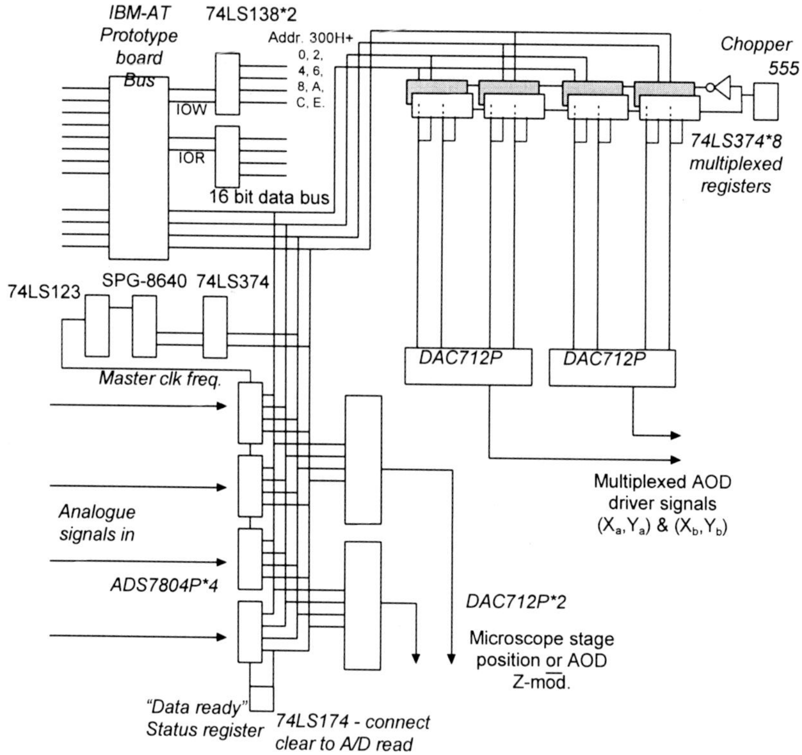

Fig. 6 There are four 12-bit, A/D converters and four 16-bit, D/A converters, two of which are controlled by two multiplexed sets of 16-bit registers (4 × 2 sets of 74LS374). These registers are soldered one on top of the other with the exception of the chip select pins (which are connected to different address lines) and the output enable pins (which are chopped using a 555 timer that runs with an even duty cycle at 10 kHz). Timing of A/D conversions and D/A conversions is clocked by a 6-bit programmable oscillator. The oscillator clock pulse initiates conversion on all four A/D chips and after transition of the Busy bit (stored in a 4-bit register; 74LS173). Valid data can be read in, and the read signal itself resets the Busy bit until the next conversion is complete.

The interferometer design used by Svoboda and Block (1994a) is not easily compatible with a dual trapping arrangement because the signal is derived from the displacement of the object within the trapping beam. If multiple optical traps are used, there will be multiple signals. However, if the traps are produced by the scanning method presented here, the position signals can be demodulated to extract the position of each object in each trap's position. We have not made such a detector yet, but this principle could be useful if optical trapping mechanical measurements are to be made simultaneously with low-light fluorescence measurements (which are incompatible with the bright-field imaging system used for four-quadrant position detectors).

Modulation of the x, y, or z inputs to the AOD produces "sum and difference" diffraction spots that create "ghost" traps. These will cause the main trap to be broadened or, if the modulation frequency is high enough, they can be used to generate extra traps. For example, the 10-kHz modulation (and 35-MHz center frequency) used to synthesise our dual trap produces reflections at 35,010,000 and 34,990,000 Hz. This will produce extra traps, 20 nm on either side of the main trap. Because these are rather small spacings, the effect will simply be to broaden the main trap. However, if the Z modulation input is set at 5 MHz, then extra reflections will be produced that can be used as independently controllable optical traps (in fact three traps will be produced altogether). This technique would allow multiple traps to be synthesized on the scanner by a rather different mechanism to direct chopping of the laser beam and would work at high enough frequencies (MHz) to show no chopping-induced motion at all.

Another simple method that can be used to produce two traps (one fixed, one movable) is to use a single-axis AOD scanner and adjust the angle of the incident beams so that there is approximately equal power in the first- and zero-order beams. The zero-order beam will be fixed, and the first-order beam would be adjustable in the x axis.

References

Finer, J. T., Simons, R. M., and Spudich, J. A. (1994). Single myosin molecule mechanics: Piconewton forces and nanometre steps. *Nature* **368,** 113–119.

Duncan, R. (1986). *The Advanced MS-DOS Microsoft guide.* Washington D.C.: Microsoft Press.

Molloy, J. E., Burns, J. E., Kendrick-Jones, J., Tregear, R. T., and White, D. C. S. (1995a). Movement and force produced by a single myosin head. *Nature* **378,** 209–212.

Molloy, J. E., Burns, J. E., Sparrow, J. C., Tregear, R. T., Kendrick-Jones, J., and White, D. C. S. (1995b). Single molecule mechanics of HMM and S1 interacting with rabbit or *Drosophila* actins using optical tweezers. *Biophys. J.* **68,** 298s–305s.

Molloy, J. E., and White, D. C. S. (1997). Smooth and skeletal muscle single-molecule mechanical experiments. *Biophys. J.* **72,** 984–986.

Simmons, R. M., Finer, J. T., Chu, S., and Spudich, J. A. (1996). Quantitative measurements of force and displacement using an optical trap. *Biophys. J.* **70,** 1813–1822.

Svoboda, K., and Block, S. M. (1994a). Force and velocity measured for single kinesin molecules. *Cell* **77,** 773–784.

Svoboda, K., and Block, S. M. (1994b). Biological applications of optical forces. *Annu. Rev. Biophys. Biomol. Struct.* **23,** 247–285.

Veigel, C., Bartoo, M. L., White, D. C. S., Sparrow, J. C., and Molloy, J. E. Myosin stiffness determined with an optical tweezers transducer. *Biophys. J.* Submitted.

Visscher, K., Brakenhoff, G. J., and Krol, J. J. (1993). Micromanipulation by "multiple" optical traps created by a single, fast scanning, trap integrated with the bilateral confocal scanning microscope. *Cytometry* **14,** 105–114.

INDEX

M

N

O

VOLUMES IN SERIES

Founding Series Editor
DAVID M. PRESCOTT

Volume 1 (1964)
Methods in Cell Physiology
Edited by David M. Prescott

Volume 2 (1966)
Methods in Cell Physiology
Edited by David M. Prescott

Volume 3 (1968)
Methods in Cell Physiology
Edited by David M. Prescott

Volume 4 (1970)
Methods in Cell Physiology
Edited by David M. Prescott

Volume 5 (1972)
Methods in Cell Physiology
Edited by David M. Prescott

Volume 6 (1973)
Methods in Cell Physiology
Edited by David M. Prescott

Volume 7 (1973)
Methods in Cell Biology
Edited by David M. Prescott

Volume 8 (1974)
Methods in Cell Biology
Edited by David M. Prescott

Volume 9 (1975)
Methods in Cell Biology
Edited by David M. Prescott

Volume 10 (1975)
Methods in Cell Biology
Edited by David M. Prescott

Volume 11 (1975)
Yeast Cells
Edited by David M. Prescott

Volume 12 (1975)
Yeast Cells
Edited by David M. Prescott

Volume 13 (1976)
Methods in Cell Biology
Edited by David M. Prescott

Volume 14 (1976)
Methods in Cell Biology
Edited by David M. Prescott

Volume 15 (1977)
Methods in Cell Biology
Edited by David M. Prescott

Volume 16 (1977)
Chromatin and Chromosomal Protein Research I
Edited by Gary Stein, Janet Stein, and Lewis J. Kleinsmith

Volume 17 (1978)
Chromatin and Chromosomal Protein Research II
Edited by Gary Stein, Janet Stein, and Lewis J. Kleinsmith

Volume 18 (1978)
Chromatin and Chromosomal Protein Research III
Edited by Gary Stein, Janet Stein, and Lewis J. Kleinsmith

Volume 19 (1978)
Chromatin and Chromosomal Protein Research IV
Edited by Gary Stein, Janet Stein, and Lewis J. Kleinsmith

Volume 20 (1978)
Methods in Cell Biology
Edited by David M. Prescott

Advisory Board Chairman
KEITH R. PORTER

Volume 21A (1980)
Normal Human Tissue and Cell Culture, Part A: Respiratory, Cardiovascular, and Integumentary Systems
Edited by Curtis C. Harris, Benjamin F. Trump, and Gary D. Stoner

Volume 21B (1980)
Normal Human Tissue and Cell Culture, Part B: Endocrine, Urogenital, and Gastrointestinal Systems
Edited by Curtis C. Harris, Benjamin F. Trump, and Gary D. Stoner

Volume 22 (1981)
Three-Dimensional Ultrastructure in Biology
Edited by James N. Turner

Volume 23 (1981)
Basic Mechanisms of Cellular Secretion
Edited by Arthur R. Hand and Constance Oliver

Volume 24 (1982)
The Cytoskeleton, Part A: Cytoskeletal Proteins, Isolation and Characterization
Edited by Leslie Wilson

Volume 25 (1982)
The Cytoskeleton, Part B: Biological Systems and *in Vitro* Models
Edited by Leslie Wilson

Volume 26 (1982)
Prenatal Diagnosis: Cell Biological Approaches
Edited by Samuel A. Latt and Gretchen J. Darlington

Series Editor
LESLIE WILSON

Volume 27 (1986)
Echinoderm Gametes and Embryos
Edited by Thomas E. Schroeder

Volume 28 (1987)
***Dictyostelium discoideum:* Molecular Approaches to Cell Biology**
Edited by James A. Spudich

Volume 29 (1989)
Fluorescence Microscopy of Living Cells in Culture, Part A: Fluorescent Analogs, Labeling Cells, and Basic Microscopy
Edited by Yu-Li Wang and D. Lansing Taylor

Volume 30 (1989)
Fluorescence Microscopy of Living Cells in Culture, Part B: Quantitative Fluorescence Microscopy—Imaging and Spectroscopy
Edited by D. Lansing Taylor and Yu-Li Wang

Volume 31 (1989)
Vesicular Transport, Part A
Edited by Alan M. Tartakoff

Volume 32 (1989)
Vesicular Transport, Part B
Edited by Alan M. Tartakoff

Volume 33 (1990)
Flow Cytometry
Edited by Zbigniew Darzynkiewicz and Harry A. Crissman

Volume 34 (1991)
Vectorial Transport of Proteins into and across Membranes
Edited by Alan M. Tartakoff

Selected from Volumes 31, 32, and 34 (1991)
Laboratory Methods for Vesicular and Vectorial Transport
Edited by Alan M. Tartakoff

Volume 35 (1991)
Functional Organization of the Nucleus: A Laboratory Guide
Edited by Barbara A. Hamkalo and Sarah C. R. Elgin

Volume 36 (1991)
***Xenopus laevis:* Practical Uses in Cell and Molecular Biology**
Edited by Brian K. Kay and H. Benjamin Peng

Series Editors
LESLIE WILSON AND PAUL MATSUDAIRA

Volume 37 (1993)
Antibodies in Cell Biology
Edited by David J. Asai

Volume 38 (1993)
Cell Biological Applications of Confocal Microscopy
Edited by Brian Matsumoto

Volume 39 (1993)
Motility Assays for Motor Proteins
Edited by Jonathan M. Scholey

Volume 40 (1994)
A Practical Guide to the Study of Calcium in Living Cells
Edited by Richard Nuccitelli

Volume 41 (1994)
Flow Cytometry, Second Edition, Part A
Edited by Zbigniew Darzynkiewicz, J. Paul Robinson, and Harry A. Crissman

Volume 42 (1994)
Flow Cytometry, Second Edition, Part B
Edited by Zbigniew Darzynkiewicz, J. Paul Robinson, and Harry A. Crissman

Volume 43 (1994)
Protein Expression in Animal Cells
Edited by Michael G. Roth

Volume 44 (1994)
***Drosophila melanogaster:* Practical Uses in Cell and Molecular Biology**
Edited by Lawrence S. B. Goldstein, and Eric A. Fyrberg

Volume 45 (1994)
Microbes as Tools for Cell Biology
Edited by David G. Russell

Volume 46 (1995)
Cell Death
Edited by Lawrence M. Schwartz, and Barbara A. Osborne

Volume 47 (1995)
Cilia and Flagella
Edited by William Dentler, and George Witman

Volume 48 (1995)
***Caenorhabditis elegans:* Modern Biological Analysis of an Organism**
Edited by Henry F. Epstein, and Diane C. Shakes

Volume 49 (1995)
Methods in Plant Cell Biology, Part A
Edited by David W. Galbraith, Hans J. Bohnert, and Don P. Bourque

Volume 50 (1995)
Methods in Plant Cell Biology, Part B
Edited by David W. Galbraith, Don P. Bourque, and Hans J. Bohnert

Volume 51 (1996)
Methods in Avian Embryology
Edited by Marianne Bronner-Fraser

Volume 52 (1997)
Methods in Muscle Biology
Edited by Charles P. Emerson and H. Lee Sweeney

Volume 53 (1998)
Nuclear Structure and Function
Edited by Miguel Berrios

Volume 54 (1998)
Cumulative Subject Index
Volumes 32–53

ISBN 0-12-564157-5
90018
9 780125 641579